get a grip on

GENETICS

Get a Grip on GENETICS

MARTIN BROOKES

TIME® LIFE BOOKS

Time-Life Books is a division of Time Life Inc.

TIME LIFE INC.
President and CEO: George Artandi

TIME-LIFE BOOKS
President: Stephen R. Frary

TIME-LIFE CUSTOM PUBLISHING

Vice President and Publisher	Terry Newell
Vice President of Sales and Marketing	Neil Levin
Project Manager	Jennie Halfant
Director of Acquisitions	Jennifer Pearce
Director of Design	Christopher M. Register
Director of Special Markets	Liz Ziehl

This book was conceived, designed, and produced by The Ivy Press Limited,
2-3 St. Andrews Place, Lewes, East Sussex BN7 1UP, England

Art Director	Peter Bridgewater
Editorial Director	Sophie Collins
Designer	Angela Neal
Commissioning Editor	Andrew Kirk
Picture Research	Vanessa Fletcher
Illustrations	Andrew Kaulman

First Printing.
Reproduction and printing in Hong Kong by Hong Kong Graphic and Printing Ltd.,

LIBRARY OF CONGRESS CATALOGING-IN-PUBLICATION DATA
Brookes, Martin, 1967-
Genetics / Martin Brookes.
p. cm.--(Get a grip on ....)
Includes index
ISBN 0-7370-0035-X
1. Genetics. I. Title II. Series
QH430.B77 1998
576.5--dc21 98-28745
CIP

CONTENTS

INTRODUCTION

VARIETY IS THE SPICE OF LIFE

our genes link us to the past and future

* Next time you are standing in front of a mirror, take a long hard look at yourself. What do you see? Blond hair, blue eyes? Black hair, brown eyes? No hair at all? Look a bit closer. How long is your nose? What shape are your ear lobes? How much do you weigh? What color is your skin? How hairy are you? Are you male or female?

WHAT IS GENETICS?

Genetics is the branch of biology concerned with the study of **heredity** (the transmission of characteristics from parents to their offspring) and **variation** (the differences observable between all living things).

THE ONE AND ONLY

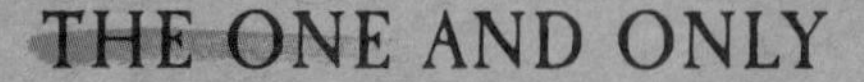

* No two people are the same. Even identical twins, who share identical genetic recipes, are not exactly alike. This uniqueness comes partly from our GENES, the set of coded instructions within our bodies. Genes are like a recipe, a set of guidelines that help to determine the way the body works and its physical appearance. ***Your genes are inherited from your parents, one half coming from your mother and the other half from your father.*** That is why we often resemble our

families share similar genetic recipes

parents and brothers or sisters more than anyone else—because we have similar GENETIC RECIPES. Our genes connect us not only with our immediate ancestors (our parents, grandparents, and so on), but also with our evolutionary relatives. Humans, chimpanzees, aardvarks, oak trees, bacteria, and the millions of other species on the planet all share a common genetic heritage.

some people are tall and slim...

Why is everyone unique?

Our uniqueness doesn't just come from our genes. Our environment (including diet, exercise, education, and a unique set of social and family experiences) is also important in making us what we are.

HISTORY REVEALED

* Modern genetics is making it possible to explore that genetic heritage, to discover the previously hidden recipes within ourselves and all living things. Sometimes exciting, sometimes terrifying, a genetic recipe can read like a book of revelations.
* It can reveal our biological history; it can tell us why we are how we are; and, to some extent, it can even predict our future.

some people are short and fat

SUPERSTITIOUS SIMILARITIES

* Genetics is one of the newest of all the sciences. Most of its significant discoveries belong to the 20th century. But humans have always been fascinated by inheritance. Before any scientific understanding emerged, there was an entire cottage industry of myth and superstition.

Where Does the term "Gene" Come From?

The term "gene" was coined in 1909 by Danish biologist ***Wilhelm Johannsen*** *(1857–1927) to describe a unit of heredity.*

the power of suggestion

GOOD IMPRESSIONS

* One of the most common beliefs, found in all cultures, was that the impressions of an expectant mother could influence the characteristics of her unborn child. In ancient Greece, for example, pregnant women were advised to spend their time gazing at noble statues and other objects of beauty if they wanted their children to be born beautiful. In 19th-century France, the fine paintings in collections such as the Louvre were the recommended treatment for mothers-to-be.

BAD IMPRESSIONS

* As well as being encouraged into certain behaviors thought to have a positive effect, pregnant women were also discouraged from any activities

believed to have potentially damaging consequences for the unborn child. In East Anglia, in England, for example, women refrained from eating strawberries when pregnant, for fear that it would cause a strawberry-colored birthmark on their child's skin.

noise pollution

* During the Second World War, the birth of mentally retarded children in East Anglia was held to be a consequence of the traumas suffered by expectant mothers during bombing raids. Even today, the belief persists that spilling tea or coffee over a woman's stomach during pregnancy will cause birthmarks.

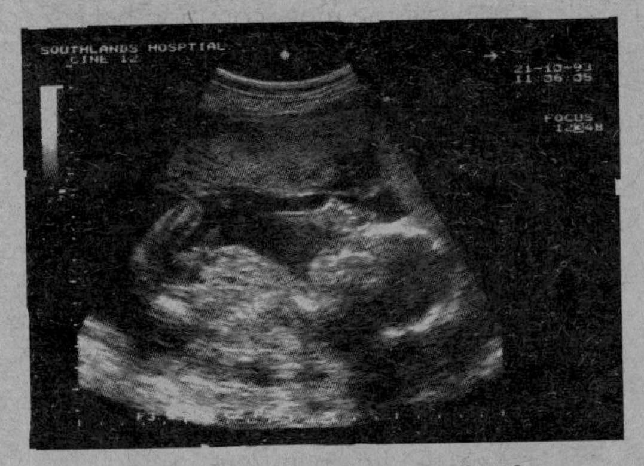

a mother's lifestyle affects the fetus

* Of course, nowadays it is well known that a mother's behavior during pregnancy can affect the health of her unborn child, since the hazards of tobacco and other drugs are widely attested. But there are good scientific reasons why this should be, and such effects have nothing to do with the myth of maternal impressions.

The Elephant Man

Joseph Merrick (1860–1890) was in no doubt about the influence of maternal impressions. Shortly before he was born, Merrick's mother had been jostled and knocked down by a circus elephant. Merrick was convinced that this was the cause of his own extreme disfigurement. Merrick attributed the rarity of his condition to the rarity of rogue elephants in rural England. In fact, Merrick was almost certainly suffering from a very rare genetic disorder known as Proteus syndrome, which causes gross deformation of bone and other tissues.

GENE GEOGRAPHY

* Today, we recognize that inheritance has a material basis. The myth and folklore of old has been replaced by modern scientific explanations of differences and similarities. Genes are real physical things residing within the cells of our bodies.

HOW MANY CHROMOSOMES?

Species	Number of Chromosomes
Human	46
Chimpanzee	48
Dog	78
Horse	64
Fruit fly	8
Pea	14

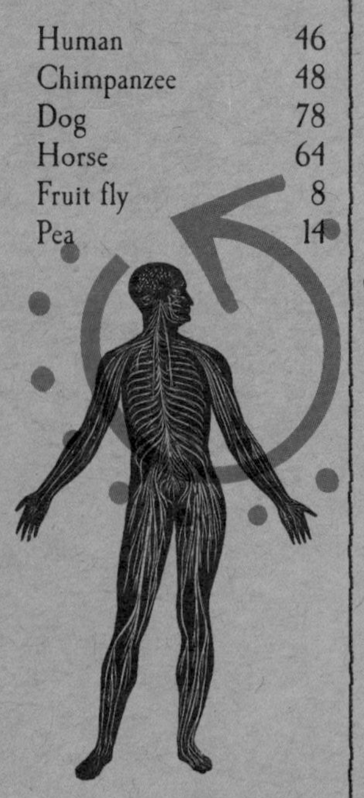

a human has 46 chromosomes

DESIGN FOR LIFE

* All living things are made up of microscopic compartments, known as CELLS. Some organisms, such as bacteria and amoebas, consist of just a single cell. But ***most organisms, including humans, are made up of billions of individual cells.***

TISSUES AND ORGANS

* Groups of cells form tissues that are specialized to perform different functions. For example, skin cells give the body external protection; muscle cells enable movement; bone cells give support; red blood cells transport oxygen from the lungs to other tissues around the body; and white blood cells protect the body against infection. Tissues, in turn, are grouped together to form the distinct organs of the body, such as the heart, lungs, liver, and kidneys.

SIZE AND SHAPE

* Although cells vary enormously in size and function, they all have the same basic design. Within each cell, there is an inner compartment called the NUCLEUS, which contains long, thin threadlike structures called CHROMOSOMES. ***Almost every cell in the body carries an identical set of chromosomes. Genes are arranged linearly on the chromosomes, like beads on a necklace.***

a necklace of genes

HOW MANY?

* The number of chromosomes differs between the different species of animals and plants—but it is always an even number, because ***chromosomes come in pairs***. One member of each pair comes from the mother, and the other from the father. However, it is not so much the number of chromosomes that makes a species unique, but the specific genes that are carried on the chromosomes.

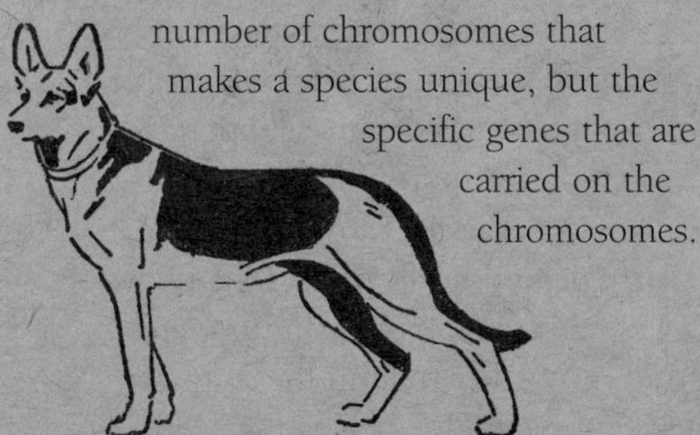

a dog has 78 chromosomes

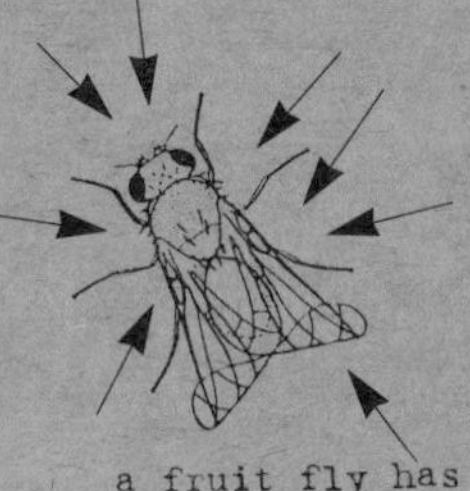

a fruit fly has 8 chromosomes

KEY WORDS

CELL: the structural and functional unit of life

NUCLEUS: an inner compartment within the cell that contains the chromosomes; found in all but the simplest organisms

CHROMOSOMES: threadlike structures in the cell nucleus that contain the genes

HOMOLOGOUS CHROMOSOMES: chromosomes that are similar in size and genetic constitution; one member of each pair of homologous chromosomes comes from the mother, and the other from the father

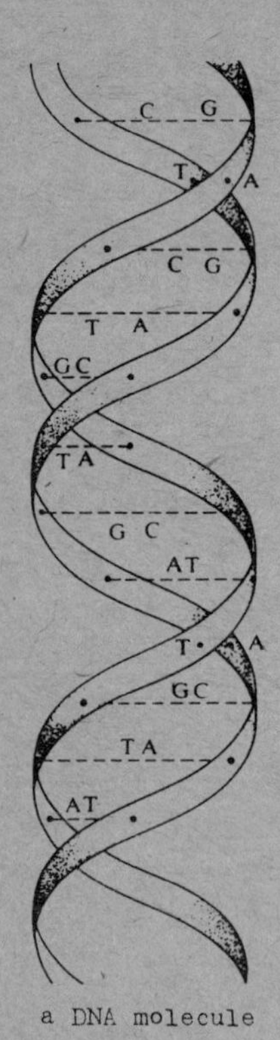

a DNA molecule

ONE LONG MOLECULE

* Deoxyribonucleic acid may well be the longest term in this book—but mercifully most people use its more familiar acronym, DNA, when referring to this important substance. DNA is the genetic material: it's what genes are made of, and its structure holds the key to how the genetic code works.

SPIRALING SIMPLICITY

* There is a lot of mystique surrounding DNA. It's a remarkable molecule, but when all's said and done, it is just that—a molecule made up of atoms held together by chemical bonds. As molecules go, however, it is extremely long. If you stretched out the DNA from a single human cell, it would measure about 5 feet (1.5m) in length, which is about 100,000 times longer than the cell itself! Not surprisingly, the DNA has to be very tightly folded to fit into the cell nucleus.

* ***The shape of the DNA molecule is one of its most characteristic features. It resembles a spiral staircase—or*** DOUBLE HELIX, ***to give it its proper scientific name. The molecule has two intertwining*** STRANDS ***with links across the middle, which form the "stairs."***

* Despite its length, DNA is actually quite a simple molecule. Each strand

WHAT IS DNA?

DNA (Deoxyribo-Nucleic Acid) is the genetic material of most living organisms. Each chromosome is effectively one extremely long molecule of DNA. The chemical building blocks of DNA are nucleotides. A nucleotide consists of a sugar molecule and a phosphate molecule bound to one of the four DNA bases—**Thymine, Adenine, Guanine,** and **Cytosine.**

consists of a linear arrangement of basic chemical units called NUCLEOTIDES. These come in four different varieties, which can be represented by the letters A, G, C, AND T. A single strand of DNA can have any sequence of these four letters. For example, here is an arbitrary bit of DNA sequence:

AATCATTCGGTACG

the letters on the DNA molecule always come in pairs

Given the sequence of letters on one DNA strand, we know instantly what the sequence of letters on the other strand must be, because the letters form complementary pairings. A always pairs with T, and G always pairs with C. For the example given above, the complementary sequence would be:

TTAGTAAGCCATGC

★ In effect, a DNA molecule is simply two extremely long sequences of complementary letters.

DNA Facts and Figures

- *If the DNA from all the cells in the human body was stretched out and laid end to end, it would reach to the moon and back—a distance of more than 435,000 miles (700,000km).*
- *The DNA from a single human cell contains about 3 billion "letters."*
- *The DNA molecule is a right-handed helix. In other words, it coils in a clockwise direction.*

THE FOUR-LETTER CODE

★ Genes themselves do not directly affect the way people look. It is their coded products—proteins—that exert their effects on the body. Proteins are the chemical building blocks that give structure to living things. They come in all shapes and sizes.

proteins are the bricks and mortar of life

How many genes?

The number of genes varies between different species. Simple organisms like bacteria may have as few as 2,000 while humans have about 100,000 genes

BRICKS AND MORTAR

★ If a human being was a house, then the bricks and mortar would represent its proteins. Proteins can also be ENZYMES—biological catalysts that increase the rate of chemical reactions in the body. ***The human body produces thousands of different proteins, all "coded for" by the four-letter language of DNA.***

WHAT'S IN A PROTEIN?

★ Proteins are themselves made up of chains of smaller chemical building blocks called amino acids. AMINO ACIDS come in 20 different varieties, and each has its own unique chemical properties.

★ The chemical properties of a protein are determined not only by its specific amino-acid sequence, but by the way in which the chain of amino acids is twisted and folded into a characteristic and complex three-dimensional shape.

AS EASY AS AGCT...?

* So how does the DNA CODE actually work? Well, unlike the English alphabet, which has 26 letters, the DNA alphabet has only four: A, G, C, and T. But just as the information in the previous sentence was determined by the sequence of the 26 possible letters, so the information in DNA is determined by the sequence of its four letters.

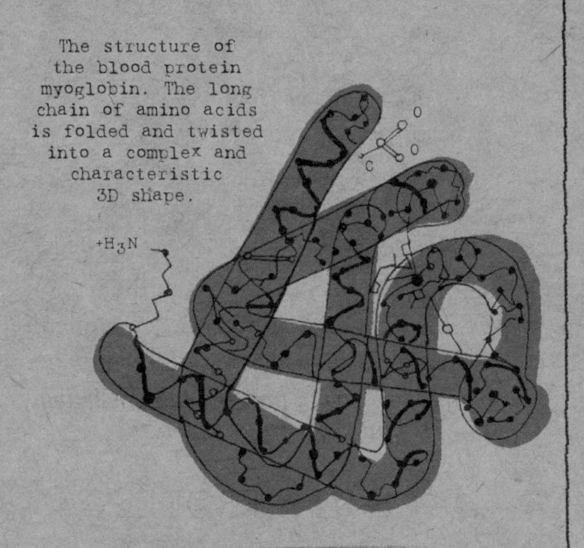

The structure of the blood protein myoglobin. The long chain of amino acids is folded and twisted into a complex and characteristic 3D shape.

* ***A DNA sequence is actually a string of three-letter "words," each word being the code for an amino acid***. So the sequence AGCTTCCGATCGGTA, for example, would actually read as AGC TTC CGA TCG GTA. Each triplet of letters specifies one of the 20 amino acids. For example, the triplet AGC is the code for the amino acid serine; TTC is the code for the amino acid known as phenylalanine; and so on.

UNIVERSAL LIFE

DNA is a universal code. A triplet of DNA letters always code for the same amino acid—no matter whether it appears in the genetic recipe of a bacterium, a cabbage, an aardvark, a human being, or any other species.

DNA PUNCTUATION

Like human languages, the language of DNA has punctuation marks to break up and clarify information. Specific triplet sequences act as stop signs, marking the boundaries where one gene ends and another begins. Therefore, a gene is the total sequence of letters that codes for a specific protein. Most genes are made up of at least 1,000 letters.

DIVIDING CELLS

*** We all start life as a single cell, created by a sperm from the father fusing with an egg from the mother.**

Billions

The first cell of a new life will divide by *mitosis* to produce a final count of billions. Each of these cells will contain a genetic recipe which is identical to the one present at conception.

AFTER FERTILIZATION

* After fertilization, the single-celled embryo begins to grow and divide. One cell splits in two to form two new identical "daughter" cells. Each of these cells then divides to become four new cells. The four cells then become eight cells, the eight cells become sixteen, and so on. Every time a cell divides, a copy of its chromosomes is made, so that each of the two new daughter cells bears a replica set. This cell division is called MITOSIS.
* The first cell of a new life will divide by mitosis to produce a final count of billions. Each of these cells will contain a genetic recipe identical to the one present at conception.

MITOSIS IN WORDS AND PICTURES

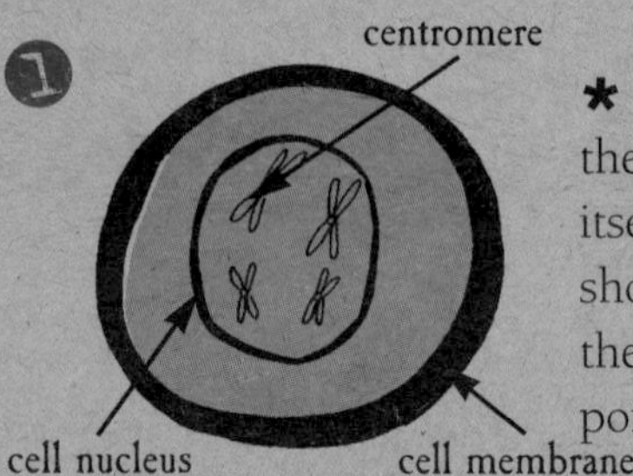

* Before a cell divides, each one of the 46 chromosomes makes a copy of itself. Only four chromosomes are shown here for clarity. The original and the copy are joined together at a central point called the CENTROMERE.

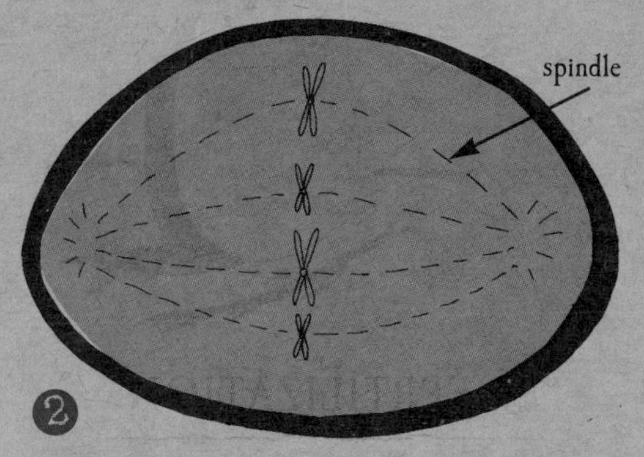

The MEMBRANE of the nucleus dissolves, and a SPINDLE of protein fibers forms on which the chromosomes line up.

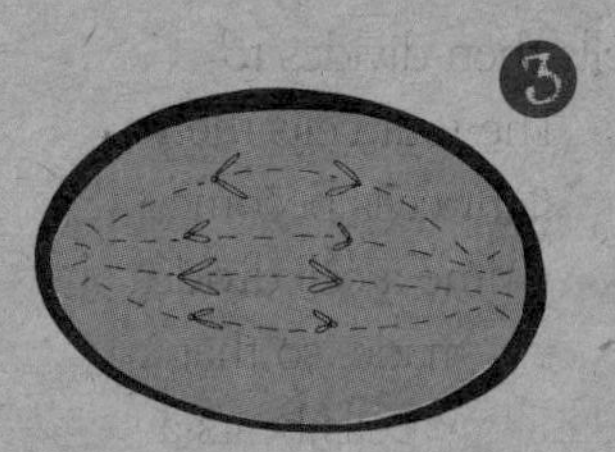

The spindle fibers then pull apart each chromosome and its copy to opposite ends of the cell.

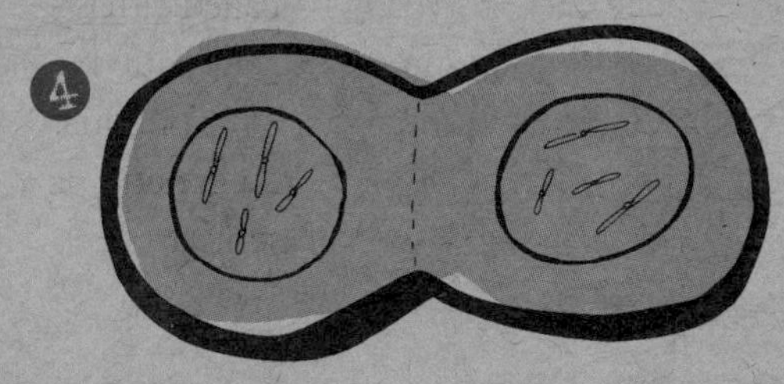

The spindle fibers dissolve, and membranes form around each of the sets of chromosomes. And finally... the cell divides in two.

KEY WORDS

FERTILIZATION: the fusion of a male sperm cell with a female egg cell to form the first cell of a new life

EMBRYO: a young organism, before it emerges from an egg, a seed, or its mother; before birth, a developing child is known as an embryo for the first two months of life and is then referred to as a fetus

MITOSIS: a type of cell division that results in two daughter cells, each carrying an identical set of chromosomes to each other and to their parent cell

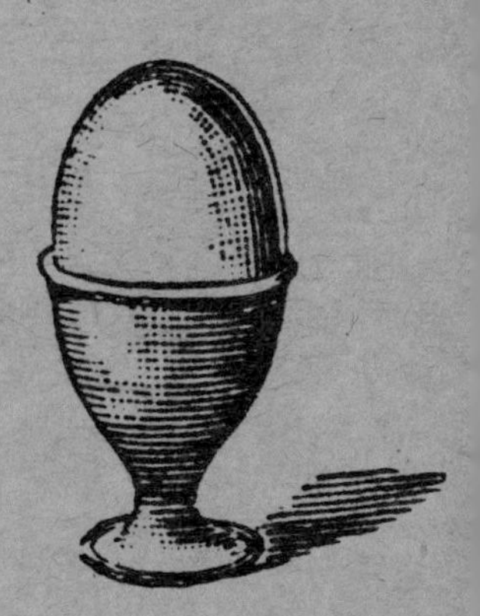

MAKING COPIES

* How does a chromosome make a copy of itself at the beginning of mitosis? There is no magic wand. A chromosome is effectively one long sequence of DNA. And DNA is a self-replicating molecule.

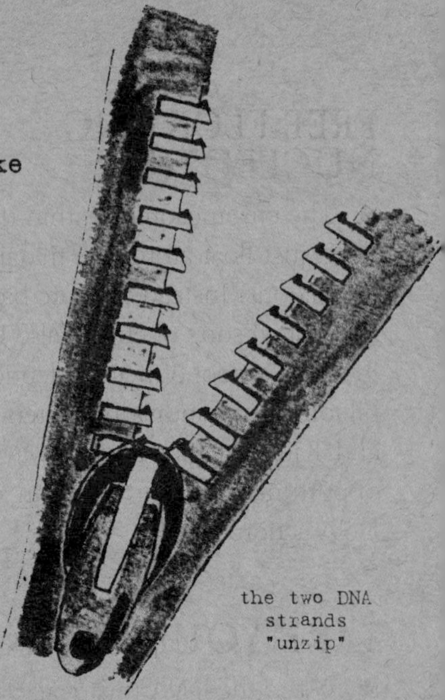

the two DNA strands "unzip"

UNZIPPING

* DNA is made up of two intertwining strands of complementary letters, held together by chemical bonds. Before copying starts, these bonds weaken and the two strands loosen their grip on one another. ***The DNA literally unzips itself, leaving the two strands exposed.*** Each of these strands becomes a template for its own replication.

duplicate copies are made of each chromosome

HELPFUL ENZYMES

DNA replication is aided and abetted by a family of enzymes known as **DNA polymerases**. At the start of replication, these enzymes anchor themselves to the tip of each chromosome and "supervise" the addition of free DNA letters to the DNA template, making sure that the crucial complementarity of the pairings is maintained.

FREE-FLOATING NUCLEOTIDES

★ The chromosomes within the nucleus don't just float around in midair like a lot of balloons. Instead they are bathed in a soup of all sorts of chemicals. This soup includes a pool of free nucleotides—billions of the individual letters A, G, C, and T just hanging around waiting for an opportunity to get involved in a bit of DNA action.

Timing

Though the exact time may vary depending on the species, the cell type, and the growing conditions, on average it takes a few hours for a cell to make a complete copy of its chromosomes.

TAKE YOUR PARTNERS

★ When the DNA unzips itself, these floating letters attach themselves to each of the two newly exposed DNA strands. But they don't just attach in any old sequence. The complementarity of the letters ensures that a free letter "A" can only pair up with a "T" on the template DNA strand, and a free "G" can only pair up with a "C."

★ In this way, an entire complementary sequence of letters is formed alongside each of the two unzipped strands. When all the free nucleotides have paired up with their complementary letters on the original strands, what you are left with is two identical copies of the original DNA sequence.

chromosome soup

SPLITTING PAIRS

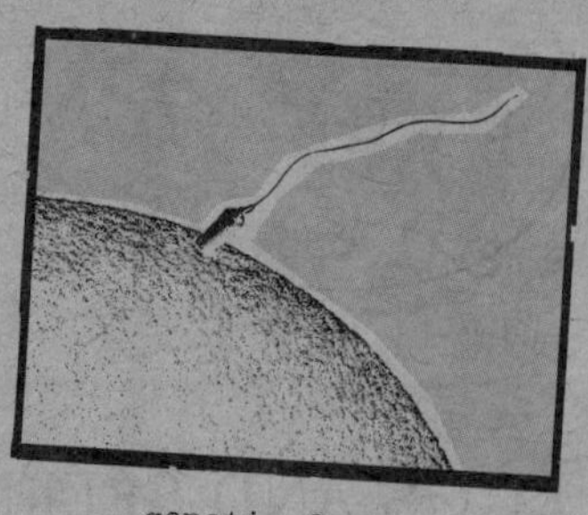

genetic fusion

★ Genes, like chromosomes, come in pairs. Each pair of chromosomes carries two copies of a gene, one copy on each chromosome. But the two genes in a gene pair are not always exact replicas of each other. Though they will encode the same basic information, such as giving color to your eyes, the precise information specified in their DNA sequence—the actual color of the eyes—may differ. It is this genetic variation that explains why there is variation—hair colors, eye colors, and so on—from one person to the next.

KEY WORDS

MEIOSIS: a type of cell division that leaves each of the two daughter cells with half the number of chromosomes contained in the parent cell; meiosis is exclusive to the germ-line cells

ALLELES: different versions of a gene are sometimes referred to as alleles

GERM-LINE CELL: a cell that divides to produce gametes

SOMATIC CELLS: "body cells"—i.e. all cells except gametes

GENETIC SEX

Every gene pair is made up of one copy from the mother and one from the father. So ***each parent transmits only half their genetic recipe to their offspring***. When a sperm fuses with an egg, half the father's chromosomes join with half the mother's chromosomes to create an embryo with a mixture of maternal and paternal genes. Sex is not just for pleasure—it's also for mixing genes!

EGGS AND SPERM

* Unlike all the other cells in the body, the sex cells—the sperm and the egg—each contain half the normal number of chromosomes, being made up of one half of each of the chromosome pairs. Halving the number of chromosomes ensures that when a sperm fuses with an egg at fertilization, the original complement of chromosomes is restored. ***In humans, for example, sperm and egg each contain 23 chromosomes. When they fuse together, we get back to the original number: 23 + 23 = 46.***

sperm are kept cool outside the body

GONAD TERRITORY

* Sex cells—known more formally as GAMETES—are produced from special cells called GERM-LINE CELLS, which are found in the testes of males and the ovaries of females. To halve the number of chromosomes, these germ-line cells undergo a special type of cell division called MEIOSIS.

Cold storage

Have you ever wondered why men's testes are situated on the outside of the body, instead of being tucked out of harm's way on the inside? Well, sperm are a bit scrawny, poor things, and ordinary body temperature (98.6°F/ 37°C) is too hot for their development. So they're stuck outside the body in the scrotum, where it's cooler.

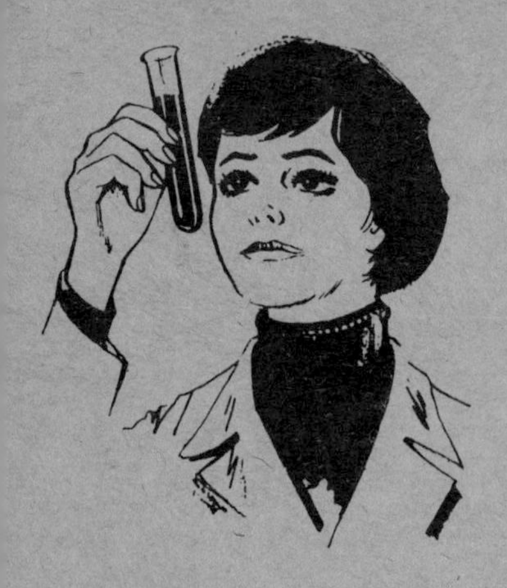

MEIOSIS IN WORDS AND PICTURES

* The early stages of meiosis are similar to those in mitosis: each chromosome makes a copy of itself, the membrane around the nucleus dissolves, and a spindle of protein fibers is formed within the cell.

HOMOLOGOUS CHROMOSOMES

1. Unlike mitosis, in meiosis the homologous chromosomes line up alongside one another on the protein fibers of the spindle.

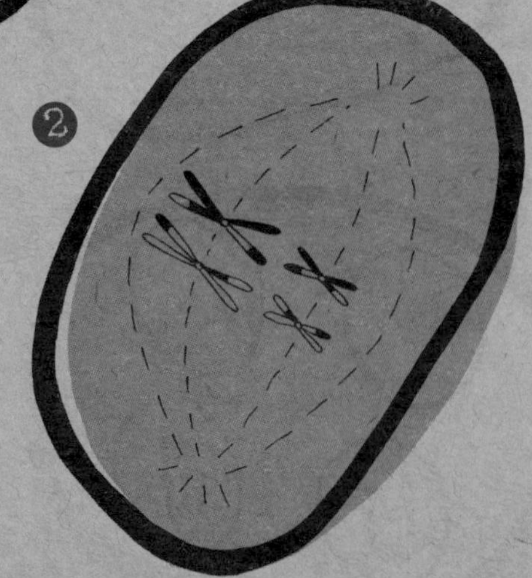

2. The homologous chromosomes then exchange segments with one another at random points along their length to create new genetic combinations on each chromosome.

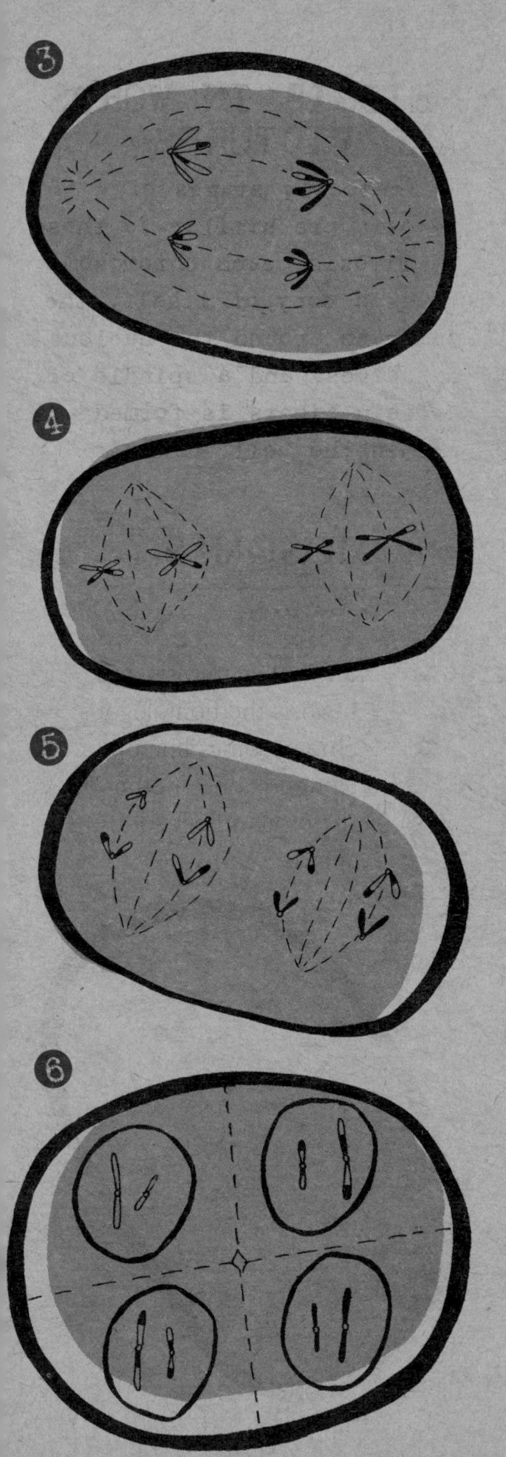

3. Each chromosome, with its new combination of genes, pulls away from its partner toward the opposite ends of the cell.

4. The spindle disperses, and two new spindles are formed at right angles. The two sets of chromosomes line up on the spindles.

5. The spindle fibers pull apart each chromosome from its copy, leaving four sets of chromosomes.

6. The spindles dissolve, and membranes form around each set of chromosomes. Finally, the cell splits into four. Each of the four new cells contains a random sample of half the genes from the original germ-line cell.

So the copy of a gene that you receive from your mother or father is a game of chance. Sexual reproduction is just a genetic lottery!

AN ORCHESTRA OF GENES

★ How a fertilized egg develops into a fully grown organism—complete with legs, arms, eyes, brain, muscles, and all the other bits and pieces—remains one of the greatest puzzles of biology. But perhaps that isn't surprising, since it's an amazingly involved and complicated process, and scientists are only just beginning to scratch the surface.

DIFFERENTIATION

★ *All the genetic information necessary for the development and* DIFFERENTIATION *of an embryo is contained in the DNA of a single fertilized egg.* In the early stages of life, the cells of an embryo remain undifferentiated. But as the cells continue to divide, they begin to take on different roles, to form the distinct tissues that will make up a fully grown organism. Since every cell contains an identical genetic recipe, how do

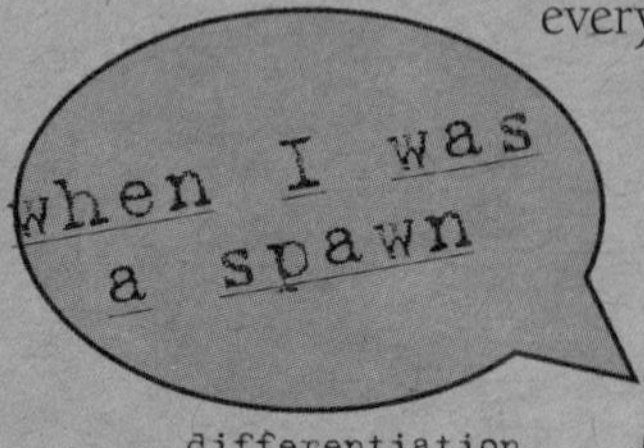

differentiation makes a frog out of a tadpole

WHAT GOES WHERE

Early on in their development, the embryos of most animals look remarkably similar. A family of genes known as the **homeobox** genes regulates the early stages of embryonic development. Amazingly, these genes seem to differ little between animals as diverse as humans and insects. Homeobox genes are like the foremen of the embryonic construction site—they specify basic positional information, such as where the head, tail, and limbs should be.

the cells "know" whether they should become a nerve cell or a muscle cell, or any of the other cell types?

* ***Genes can be turned on and off like light switches.*** Though every cell contains the genetic information necessary to produce all the proteins needed by the body, any one cell type will only use a fraction of its genetic recipe. Different genes are turned on, or EXPRESSED, in different tissues, while others remain silent.

* The cells that make up skin and hair, for example, produce lots of the protein keratin, which gives strength and protection to these tissues. But they don't produce hemoglobin, the protein that carries oxygen in the blood cells. Likewise, blood cells do not waste their time producing keratin. ***Each cell type only produces the proteins appropriate to its function.***

* But what regulates the expression of genes in different tissues? Genes of course! The sole purpose of many genes is to produce proteins that turn other genes on or off. Development is a confusing business—an incredibly complicated yet finely orchestrated system of communication, involving chemical signals coming not only from inside the body but from outside as well.

Body Builders

Biologists have established certain basics about development. It follows a programmed hierarchical pattern. In fact, it is not dissimilar to the way in which a house is built. Before cells become specialized, a basic segmented structure is laid down in the developing embryo (ground floor, first floor, roof, chimney) and the details (kitchen, bathroom, living room) are added later.

get the basics first, then concentrate on the details

MUTATIONS—THE SOURCE OF VARIETY

the cat...

* We often take the word "mutant" to mean something aberrant or deformed. But we are all mutants of one kind or another. Mutations—random genetic accidents—are the ultimate source of all genetic variation. Without mutations, there would be no variety; and without variety, there would be no evolution. If it wasn't for mutations, the earth would still be populated by a mass of identical molecules swimming around in the primordial soup.

MISTAKES

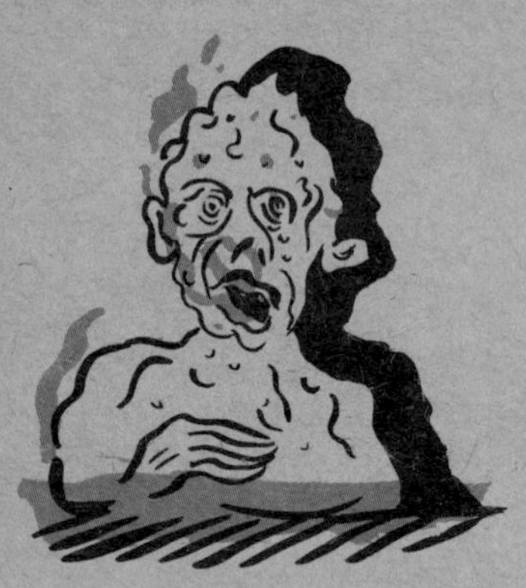

without mutations we would never have evolved

* All cells possess a suite of special ENZYMES whose job is to maintain the upkeep of DNA and ensure that it is replicated during cell division. For example, polymerase enzymes help to attach the free nucleotides to the unzipped DNA template during DNA replication; and proofreading enzymes move along newly synthesized DNA strands, making sure that all the

DIFFERENT TYPES OF MUTATION

Type of mutation	Original sequence	Mutated sequence
Substitution	ATC**G**TTAGGC	ATC**C**TTAGGC
Deletion	ATCG**TTA**GGC	ATCGGGC
Insertion	ATCGTTAGGC	ATCGT**CCA**TAGGC
Inversion	AT**CGTT**AGGC	AT**TTGC**AGGC
Duplication	AT**CGTT**AGGC	AT**CGTTCGTT**AGGC

SCRAMBLED CODE

To illustrate the ways in which mutations can affect the meaning of the DNA code, imagine a DNA sequence as a sentence written in the English language.

Original code:
THE CAT SAT ON THE MAT

Some mutations have little or no effect on the meaning, for example:
THE CAT ON THE MAT (deletion)
or
THE KAT SAT ON THE MAT
(substitution)

Others change the meaning altogether:
THE HAT SAT ON THE MAT
(substitution) or
THE CAT THE MAT (deletion)

And others make the message nonsensical:
THE CAT EHT NO TAS MAT
(inversion).

letters are positioned in their correct places. But though these enzymes are efficient, they do occasionally make mistakes, and errors can be "overlooked."

* ***Mutations—spontaneous changes in DNA—can manifest themselves in a variety of ways. They can be a substitution of one letter for another; a deletion of a single letter or blocks of letters; or an insertion, inversion, or duplication of a single letter or blocks of letters.***

* Sometimes a random change to the genetic code will result in a new version of a protein that functions better than what went before. All the genetic differences between individuals or species have come about through mutations at some point in history.

enzymes act as copy checkers

MUTATIONAL MISHAPS

Mutations can have dire consequences for an organism. Changes to the DNA sequence of a gene can translate into changes to a protein's amino-acid sequence. In extreme instances, protein production may be blocked altogether. Major mutations are rarely transmitted to subsequent generations, because they kill the organism before it has a chance to reproduce. But not all mutations have such drastic effects. Because we have two copies of any particular gene, one normal working copy may be able to "cover" for its defective partner.

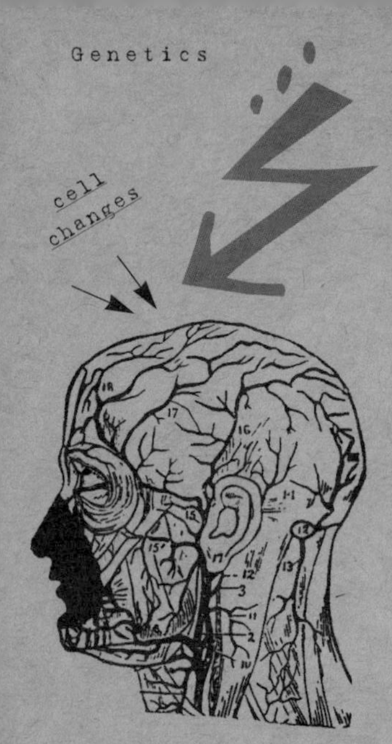

most cell changes in adults are unnoticed

DAMAGING MUTATIONS

* Mutations can occur any time, anywhere, and any place. All cells within the body are potentially at risk. But only mutations in the germ-line cells—those giving rise to the gametes—have any chance of being passed on to future generations. Of course, once a new mutation is passed on via the sperm or egg, it will become incorporated into the genetic recipe of every cell in the offspring's body. Some of the most serious human diseases are caused by these inherited mutations.

Size matters

Large genes are more vulnerable to random mutation than smaller genes. The genes responsible for the most common inherited diseases, such as cystic fibrosis and phenylketonuria, are all unusually large.

GENETIC DISEASE

* *Genetic diseases are caused by defective genes. Instead of producing a normal working protein, the mutated gene either produces an altered version of the protein which cannot function properly, or no protein at all.*

* About 5,000 genetic diseases have been described. These include SICKLE-CELL ANEMIA, CYSTIC FIBROSIS, and HEMOPHILIA, to name just a few.

SOME OF THE MOST COMMON GENETIC DISEASES

DISEASE	FREQUENCY
Sickle-cell anemia	1 in 400 (US Blacks)
Cystic fibrosis	1 in 2,000 (Whites)
Phenylketonuria	1 in 5,000 (western Europe, rare elsewhere)

SOMATIC MUTATIONS

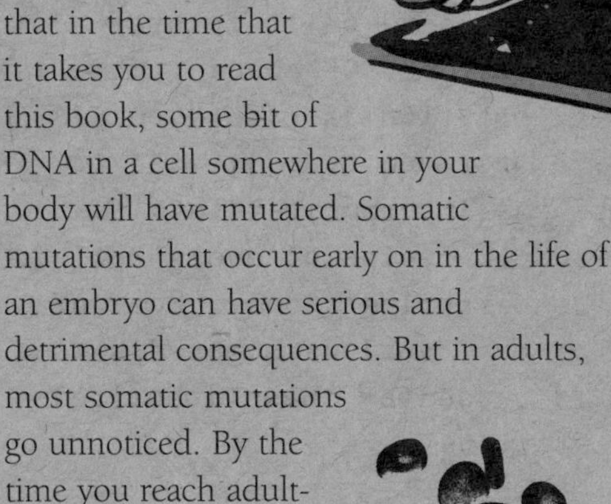

* Mutations in the somatic cells—which make up the rest of the body—are going on all the time. In fact, you can almost guarantee that in the time that it takes you to read this book, some bit of DNA in a cell somewhere in your body will have mutated. Somatic mutations that occur early on in the life of an embryo can have serious and detrimental consequences. But in adults, most somatic mutations go unnoticed. By the time you reach adulthood your body is fully developed; changes to a few cells among billions are therefore pretty insignificant.

the occasional mutated cell is insignificant

CAUSE FOR CONCERN

* However, some somatic mutations do give cause for concern. ***Cancers are the result of somatic mutations in genes that ordinarily control cell growth and division.*** Because of the mutation, a cell that has finished growing and dividing, and is now quietly going about its business, starts to divide uncontrollably, and forms a tumor.

SICKLE-CELL ANEMIA

Sickle-cell anemia was one of the first genetic diseases to be identified. It's caused by a mutation in the gene for hemoglobin, the blood protein that carries oxygen around the body. The altered version of the protein gives the red blood cells an abnormal sickle shape, which results in abnormal blood flow, blocked blood vessels, and severe anemia. Currently there is no effective treatment, and children suffering from the disease tend to live short and painful lives.

danger: mutagens

AN INVISIBLE THREAT

* Next time you light up a cigarette, bite into a bacon sandwich, go sunbathing, or stand under the mushroom cloud of a recently detonated atomic bomb, spare a thought for your DNA. Mutations aren't just the result of internal copying errors. They can also be caused by random insults from outside the body. The environment is full of agents just waiting to inflict lasting damage on your DNA.

MUTATIONAL MAYHEM

* When the Chernobyl nuclear reactor exploded in 1986, it sent a huge cloud of radioactive dust up into the atmosphere. Eventually the dust fell to earth, covering a large area of countryside. If you visit the site of the accident today, superficially everything seems normal. But the real damage lurks beneath the surface.

* *Radiation can cause gross chemical changes in DNA, throwing DNA replicating enzymes into total confusion.* Sure enough, when scientists went back to Chernobyl several years after the accident, they found evidence of mutational mayhem. Mutation rates in humans and other animals had risen dramatically, and the visible manifestations of these DNA changes were all too apparent. Many

KEY WORDS

MUTAGEN: an external agent (either a toxic chemical or radiation) capable of inducing mutations

INDUCED MUTATION: a mutation caused by a mutagen

SPONTANEOUS MUTATION: a mutation caused by a mistake during DNA replication that occurs without the influence of a mutagen

young children were suffering from thyroid cancers, and rodents were reported with spleens six times their normal size.

radiation causes mutation in all living things

THE SUNNY SIDE

* You can't escape from radiation. It's all around us. Fortunately, not all of it is as potentially damaging as the nasty stuff around Chernobyl. Ultraviolet radiation is emitted by the sun, though most of it is absorbed by the ozone layer (or what's left of it) before it has a chance to reach us.

* Cosmic radiation in outer space can also induce serious DNA damage. But unless you are planning a spacewalk in the near future, there is little to fear. What's more, help is at hand. Cells are equipped with repair enzymes that can patch up small-scale mutational damage.

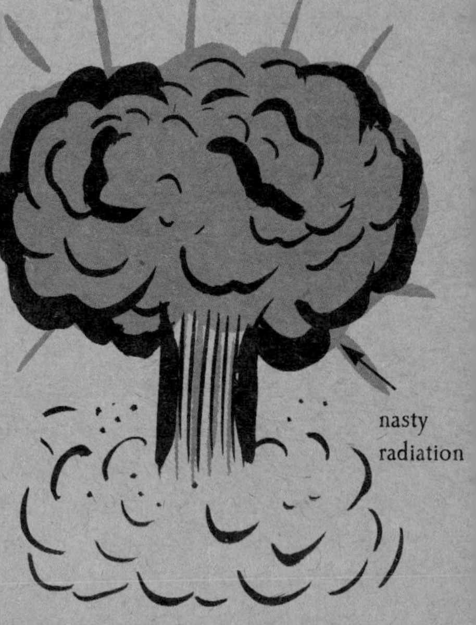

nuclear radiation is probably the most damaging

TOXIC TROUBLE

* ***Toxic chemicals can exert effects on DNA similar to those caused by radiation***. Cigarette smoke, heavy metals, some meat preservatives, and many industrial pollutants can induce panic and disorder among the letters of your genetic recipe. Many naturally occurring chemicals in food are also mutagens. Pepper and Earl Grey tea, for example, both contain natural chemical mutagens.

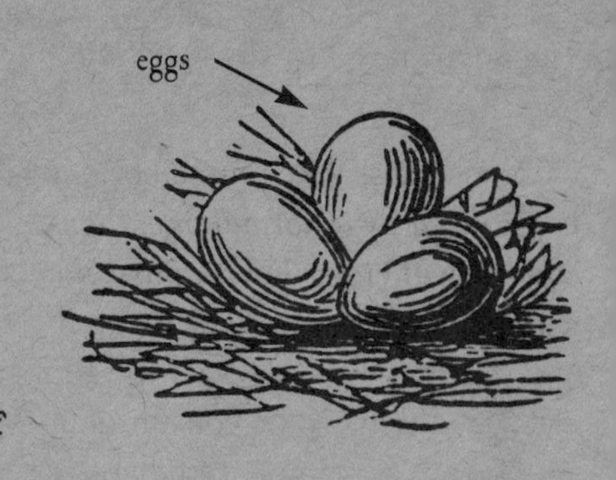

CHROMOSOME ACCIDENTS

* It's estimated that 20 percent of all pregnancies end in miscarriage. About half of these spontaneous abortions are caused by chromosomal abnormalities in the developing embryo. Genetic accidents are not just restricted to single genes. They can also extend to the whole chromosome.

KEY WORDS

CYTOGENETICS: the study of chromosomes and their inheritance

SYNDROME: a disease that causes multiple symptoms

MEIOSIS GONE WRONG

* *In humans, meiosis ensures that male sperm or female eggs contain 23 chromosom* But though this is the normal outcome of meiosis, things can occasionally go wrong. Sometimes, the chromosomes in a particular chromosome pair do not separate from one another, resulting in gametes with 22 or 24

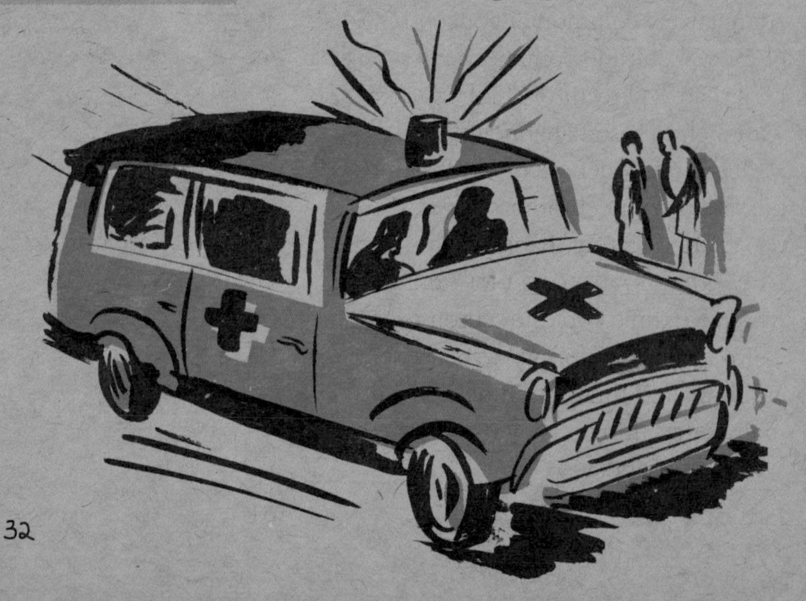

chromosomes. ***When an abnormal gamete is fertilized by a normal one, the resulting embryo will either have a missing or an extra chromosome***. As the embryo grows and divides, every new cell will bear this unusual chromosome constitution.

* Down's syndrome is perhaps the best known and most common chromosomal abnormality in humans. It occurs in about 1 in 700 live births. People with DOWN'S SYNDROME usually have 47 chromosomes, because of an extra copy of chromosome 21.

embryo development

* Although chromosomal abnormalities are relatively common at conception, most of them result in spontaneous abortion. Less than 1 percent of children are born with a chromosomal abnormality.

* The consequences of having a missing or extra chromosome are usually devastating. Babies with EDWARDS' SYNDROME, for example, who have three copies of chromosome 18, suffer from ear deformities, heart defects, and impaired muscular function, among many other things. They usually die before their first birthday.

Down's Syndrome

The risk of giving birth to a child with Down's syndrome increases with maternal age. Women aged 45 are 100 times more likely to give birth to a child that has the syndrome than women under 19. Though the exact reason is unclear, one possible explanation is related to the way in which eggs are made. A woman's entire lifetime production of eggs is effectively present at birth; consequently all the cells that will go on to divide to produce eggs are held in a suspended state until near the time of ovulation. So a woman of 45 will ovulate eggs that have been sitting dormant for 45 years—and the longer an egg sits in the ovary, the greater the chance that it will acquire a mutation.

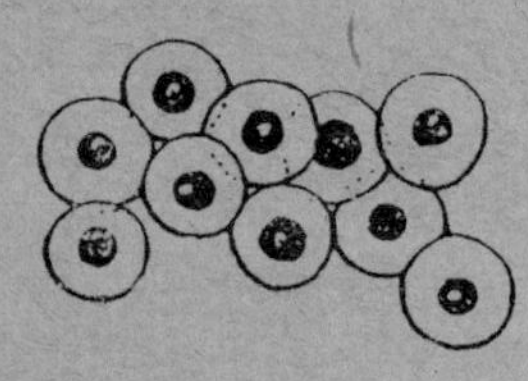

human cells have 23 pairs of chromosones

THE GENDER AGENDA

★ Humans have 23 pairs of chromosomes within each of their cells. Each chromosome resembles its partner in size, shape, and genetic constitution, with one important exception. The exception is the pair of sex chromosomes. Which combination of sex chromosomes you have determines whether you are male or female.

DETERMINING GENDER

The way that gender is determined in humans is common to many other species, but it is by no means universal. In birds and butterflies, for example, males are XX and females are XY. And in some reptiles and fish, gender is determined not by genes but by the temperature at which the fertilized eggs are incubated.

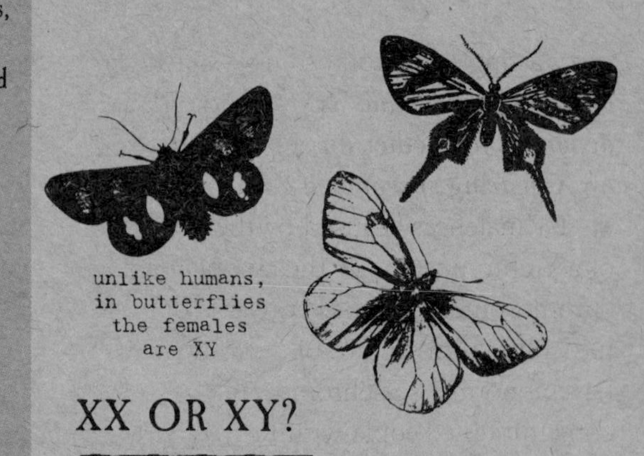

unlike humans, in butterflies the females are XY

XX OR XY?

★ In men, the sex chromosomes are made up of a large chromosome called the X CHROMOSOME, and a much smaller, stubby-looking thing known as the Y CHROMOSOME. In contrast, women have two X chromosomes.

★ Because males have two different sex chromosomes, they produce two distinct types of gametes. Half their sperm cells carry an X chromosome, and half carry a Y chromosome. In contrast, all egg cells

it's a boy!

more XY embryos are conceived than XX

carry a single X chromosome—***so whether an embryo is female (XX) or male (XY) depends on whether an egg is fertilized by an X-bearing sperm or a Y-bearing sperm***.

★ Is "maleness" caused by the presence of a Y chromosome or the absence of an X chromosome? The answer to this question has come from studies of individuals who have abnormal sex-chromosome constitutions. People with KLINEFELTER'S SYNDROME have two X chromosomes and one Y. They are fairly normal males, although they have small testes, produce little or no sperm and have rudimentary breasts. People with TURNER'S SYNDROME have just a single X chromosome. They are recognizably female, but have undeveloped internal and external genitalia and are sterile. So maleness is determined by the Y chromosome—and recent research has tracked maleness down to a single gene carried on the Y chromosome.

Sexual Inequality

One would expect boys and girls to be conceived with equal probability, but it is a fact that more boys are conceived than girls. The reason may be that Y-bearing sperm have some sort of competitive advantage as they swim up the female reproductive tract in search of eggs. More males die before birth than females; but even so, about 105 boys are born for every 100 girls.

CHAPTER 1

IN THE BEGINNING

Pop!

* The year is 1865. An Augustinian monk, Gregor Mendel, is tending a crop of common peas in the tranquil gardens of a monastery in Brünn, Austria. Among the neatly arranged rows of plants, there are recognizable and distinct varieties. Tall plants stand alongside short plants; plants with pink flowers contrast strikingly with white-flowered varieties. Splitting open the pods reveals other differences: while some plants have yellow peas, others bear the more traditional green variety.

1865—a Year to Remember

- *Civil War ends*
- *Abraham Lincoln assassinated*
- *Slavery abolished in the U.S.*
- *Joseph Lister uses antiseptics for the first time in surgical operations*
- *William Booth establishes the Salvation Army*
- *Gregor Mendel reports the results of his genetic experiments to the Brünn Society for the Study of Natural History*

PEA POWER

* When he was not in prayer, Gregor Mendel devoted himself to his pea plants. Intrigued by the way in which the DISTINCT CHARACTERISTICS of the plants were passed on from one generation to the next, Mendel spent years carefully CROSSING his pea varieties and counting the number and types of plants produced.

* Only after long and painstaking experimentation was he ready to submit the results of his studies for scientific publication. ***His work would revolutionize science***. It would also set the agenda for a century of remarkable scientific discoveries that would change the way people thought about themselves and their futures.

Mendel tending his peas

AHEAD OF HIS TIME

* Today, Gregor Mendel is hailed as the founding father of modern genetics. But the initial reaction of the scientific community was one of universal indifference instead of worldwide acclaim. His discoveries were dismissed as irrelevant because they were seemingly at odds with the prevailing ideas on inheritance.

* Mendel was simply way ahead of his time. It would be another 40 years before the importance of his work was recognized. By that time Mendel was pushing up daisies—or, more probably, peas.

THE FATHER OF GENETICS

Gregor Mendel (1822–84) entered an Augustinian monastery in Brünn, Austria (now Brno, Czechoslovakia) at the age of 21. In 1851 he was sent to Vienna University to study science, but after several failed attempts to gain a teaching qualification, resigned himself to a life of peas and celibacy. He began his plant experiments in 1856, and in 1865 reported his findings to the Brünn Society for the Study of Natural History. His "Experiments with Plant Hybrids" was published the following year. Mendel's personality, and his reputation as a monk, made him unsuited to self-publicity.

IN THE BLOOD?

* In the mid-1860s, most biologists shared the view that heredity was a process of blending. Like the mixing of paints, parental characteristics were thought to blend together in the offspring, so that a child's appearance and physical characteristics would be the average of its parents.

a mixing palette

CHARLES DARWIN

Charles Darwin (1809–82) is better known for his ideas on evolution than on genetics. He began his academic career at Edinburgh University, studying medicine. But traumatized by the blood and guts of anatomy, he relocated to Cambridge to study theology. It was there that he forged a lifelong interest in natural history, an interest that later inspired him to write *The Origin of Species*, detailing his radical new theory of evolution by natural selection.

a perfect blend

DARWIN'S THEORY

* To explain how the process of BLENDING INHERITANCE worked, **Charles Darwin** put forward a rather speculative theory of "PANGENESIS". Darwin suggested that each of the organs in the body produced particles, which he called "gemmules," that were like miniature representations of the organs to be produced in the offspring.

* It was assumed that the particles were transported in the bloodstream to the reproductive organs. At fertilization, after a sperm had fused with an egg, the gemmules from each parent would be mixed together.

turned out white again!

GARBAGE?

* Blending inheritance was consigned to the scientific garbage can, and Darwin's pangenesis theory was soon to follow. In one famous experiment, **Francis Galton** (who was Darwin's cousin) transfused blood from a black rabbit to a white rabbit. Galton reasoned that if a "gemmule" for black coloration was carried in the blood, then the offspring of the white rabbit should be black. But all the baby rabbits turned out to have white fur.

* Although Galton had disproved the pangenesis theory, the old adage "in the blood" persisted, and the expression is still widely used today.

A FATAL FLAW

In 1867, a Scottish engineer named **Fleeming Jenkin** pointed out a fatal flaw in the blending inheritance idea. He argued that blending inheritance would cause a progressive dilution of individual characteristics with each successive generation. After many generations, all individuals would look the same. But, clearly, all individuals were not the same. Differences between human beings were only too apparent, and in both animals and plants it was the differences between individuals that were central to Darwin's theory. Variety was the raw material of **natural selection**. Without it, there could be no **evolution**.

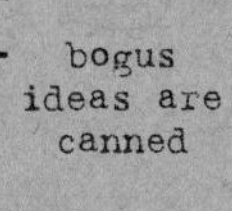

bogus ideas are canned

LAMARCKIAN INHERITANCE

if you do your exercises, your children will inherit your strength, said Lamarck

★ Another belief popular among 19th-century biologists, including Darwin, was that characteristics acquired during a person's lifetime could be inherited. If a person built up strong muscles through regular exercise, then their children would also inherit strong muscles. Historically, "the inheritance of acquired characteristics" was derived from the ideas of the French scientist Jean Baptiste de Lamarck.

JEAN BAPTISTE DE LAMARCK

Jean Baptiste de Lamarck (1744–1829) started life as a soldier, but later became interested in meteorology, chemistry, and biology. He was tremendously influential in many areas of biology and a pioneer of evolutionary thinking. Outside his native France, however, his memory has largely been eclipsed by Darwin.

THE GIRAFFE'S NECK

★ **Lamarck** believed that an organism's needs determined how it developed, and that its needs were determined by the environment in which it found itself. The effort to satisfy these "needs," he argued, could result in modifications to the organism's body that would be passed on to subsequent generations.

★ Lamarck used the now famous example of the giraffe's long neck to illustrate his idea. The ancestors of the modern giraffe would have strained their necks to feed on the leaves of the tall trees of the African savannahs. As a result, their necks would have grown a little bit longer, and an elongated neck would have been passed on to the next generation of giraffes. After many generations of stretching and straining, the result would be the long-necked animals we see today.

Lamarck

CONTROVERSY

★ By the turn of the century, the inheritance of acquired characteristics—or Lamarckian inheritance, as it is often termed—had been largely discredited. Yet the idea has resurfaced in different guises throughout this century, and even today controversy still rages over whether organisms can direct their own evolution according to their needs.

an evolutionary success

DO-IT-YOURSELF EVOLUTION

The concept of *directed mutation* is a recent reincarnation of the old Lamarckian inheritance idea. Some biologists believe that simple single-celled organisms such as bacteria may be able to direct their own evolution, though the idea remains deeply controversial.

can simple organisms direct their own evolution?

HISTORICAL FOCUS

★ Although early ideas about inheritance may seem speculative and fanciful today, a detailed understanding of inheritance was severely hampered by the limited technology of the time.

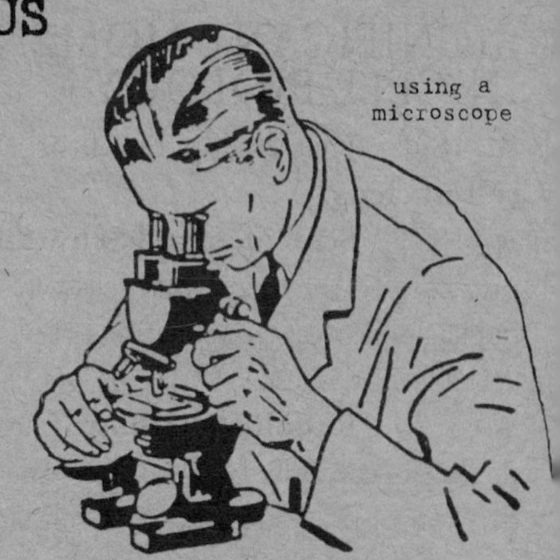

ROBERT HOOKE

The English physicist **Robert Hooke** (1635–1703) was a man of great ideas and a prolific experimenter. In 1665, he anticipated the path that biology would take 200 years later when he wrote: "The truth is, the science of Nature has been already too long made only a work of the brain and the fancy. It is now high time that it should return to the plainness and soundness of observations on material and obvious things."

MICROSCOPES

★ The discovery of the material basis of heredity came, in part, from probing the finer details of plant and animal tissues. In 1655, the physicist Robert Hooke, ***using a primitive light microscope, described small angular spaces in sections of cork***, for which he coined the term "cells." But the idea that cells were the fundamental units of all animals and plants was not confirmed until superior light microscopes became available in the early 19th century.

★ Before they could get a more REFINED PICTURE of these tiny and colorless structures, biologists had to wait until the latter part of the 19th century, when SUITABLE STAINS AND DYES were developed that could highlight and contrast the inner architecture of cells.

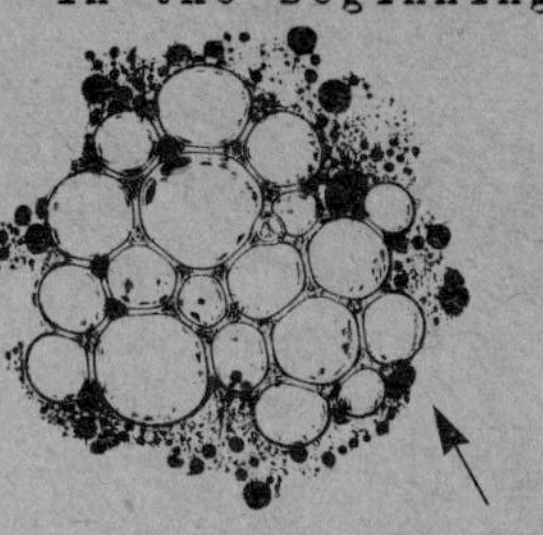

SIGNIFICANT MOMENTS IN CELL BIOLOGY

* 1833—First description of the **cell nucleus.**
* 1835—**Schleiden and Schwann** put forward the theory that cells with nuclei are the fundamental units of plant and animal tissues.
* 1855—**Rudolph Virchow** states that new cells can only be formed by the division of previously existing cells—in other words, cells cannot arise by SPONTANEOUS GENERATION.
* 1869—**Frederick Miescher** discovers NUCLEIN (DNA).
* 1879—Using new staining techniques, **Walther Flemming** identifies chromosomes in the cell nucleus (from chroma, the Greek word for color).

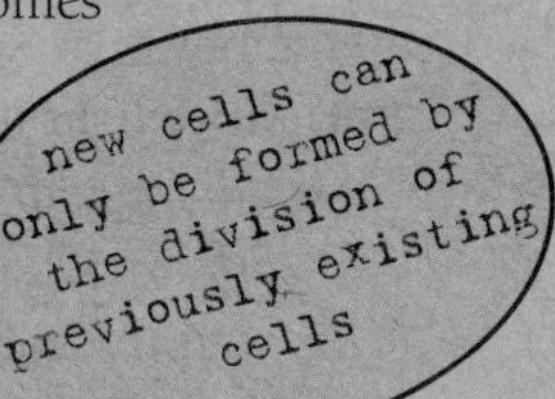

FREDERICK MIESCHER

Frederick Miescher (1844–95) was 25 years old when, in 1869, he went to Tübingen University, in Germany, to study the chemistry of white blood cells. Pus was a good source of white blood cells, so Miescher made sure he was never short of pus-soaked post-operative bandages. He found that adding acid and alkali to the cell nuclei produced a gray precipitate, previously unknown to science. Since it came from the nucleus, Miescher gave the name "nuclein" to this new organic substance. Today it is called DNA. Its discovery came 80 years before it was identified as the genetic material.

WEISMANN AND THE GERM PLASM

* As the 19th century drew to a close, a German biologist named August Weismann helped to concentrate the minds of the scientific world and steer genetics into the next century.

AUGUST WEISMANN

August Weismann (1834–1914) was a skilled microscopist, but in the mid-1880s failing eyesight forced him to concentrate more on the theoretical aspects of heredity.

‘A HEN IS ONLY AN EGG'S WAY OF MAKING ANOTHER EGG’

SAMUEL BUTLER

IN FOCUS

* **Weismann** was one of the new breed of microscopists who were bringing the internal structure and workings of cells sharply into focus. From his own observations, Weismann became convinced that the material basis of heredity was located in the chromosomes. At fertilization, the hereditary instructions from each parent were mixed together when a sperm from the father fused with the egg from the mother. The structure of the body, he believed, was therefore determined by the combination of hereditary instructions received from the two parents.

Weismann also developed a radical new theory postulating the continuity of the "germ plasm," in which he argued that the "germ plasm" (the gametes) developed independently

the hen and the egg

of the rest of the body. Moreover, the genetic recipe was passed from one generation to the next via the gametes and was unaffected by changes occurring in the rest of the body. In other words, the body was simply a vehicle for the transmission of the germ line.

* August Weismann's theory was a serious blow to the ardent followers of Lamarck. If there was a barrier between the sex cells and the rest of the body, then there was no way in which characteristics acquired during a person's lifetime could be incorporated into the genetic recipe of the germ line.

the tail cutter

MICE AT THE CUTTING EDGE

* To prove his point, Weismann carried out an experiment of virtuosic Victorian barbarity. He took a group of mice and cut off their tails. He then looked to see whether the offspring of these mice had tails that were shorter than normal. After 22 generations of cutting and copulation, he concluded that chopping off the tails of the parent mice had absolutely no influence on the tail size of their offspring.

Desperate measures

Weismann had no need to go to such gruesome lengths. Had he looked at the children of maimed soldiers, or considered the practice of circumcision, then his point would have been equally well-made.

mercy for the mice

MENDEL REVISITED

★ In the 40 years that had elapsed since Mendel's experiments on peas, the genetic landscape had undergone some radical changes. When new theories about the material basis of heredity were suddenly seen to coincide with Mendel's own ideas, it was time for Mendel to step posthumously into the limelight.

Mendel with round peas and wrinkled peas

MENDEL'S EXPERIMENTS

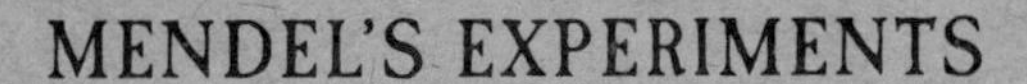

★ Mendel's genius lay in his ability to detect a regular pattern in the way that characteristics were transmitted over successive generations, and to relate these PATTERNS OF INHERITANCE to real physical phenomena.

★ Mendel began his experiments with two strains of pea plant, which were both PURE-BREEDING varieties (i.e. the offspring were always identical to the parents) but differed from each other in a single characteristic, such as pea shape. In one experiment, he pollinated the flowers of a pea plant that bore smooth, round peas with pollen from a plant that produced wrinkled ones. All the offspring from the cross had round peas. But when he crossed these round-pea plants with one another, the offspring included some plants with round peas and others with

Why Plants?

Working out the rules of inheritance depended on counting large numbers of individuals. Pea plants were ideal for this purpose. They were quick and easy to breed and, most important of all, produced lots of offspring.

evening primrose

wrinkled peas. The wrinkled characteristic had disappeared in one generation and reappeared in the next. Mendel also noticed that the two types of plants were not produced in equal numbers. There were always about three round-pea plants produced for every one wrinkled-pea plant.

THE RESULT

* Mendel realized that his results could be explained if each characteristic (i.e. pea shape) was coded for by a pair of "factors" (what we now call genes). The two genes separated from each other during the formation of gametes, so that each gamete—a male pollen grain or a female egg—carried only one of the two genes. Then at fertilization, when the pollen fused with an egg, genes united to form a new pairing.

* To explain how the wrinkled characteristic could be suppressed in one generation and reappear in the next, Mendel introduced the idea of DOMINANT AND RECESSIVE GENES. Different genes had hierarchical relationships with one another, so that when two different genes made up a pair, a dominant gene would mask the effect of a recessive gene. In the case of pea shape, round was dominant and wrinkled recessive.

DOMINANT AND RECESSIVE GENES

A **dominant gene** is one that will express its effect on a characteristic regardless of its partner gene—while a **recessive gene** is one which must be paired with an identical copy for it to express its effect.

Hugo De Vries

Mendel's work was "rediscovered" by the Dutch plant geneticist **Hugo De Vries** *(1848–1935) and others in 1900. De Vries was himself instrumental in developing early ideas about mutations in the evening primrose.*

GENETIC SYMBOLISM

* To understand Mendel's patterns of inheritance more fully, it is easiest to think (as Mendel did) of genes in terms of symbols. Take the example of the round and wrinkled peas: let's call the gene for roundness "R" and the gene for wrinkled "r."

pea pod

PEDIGREES

* Pure-breeding plants will carry two identical copies of the gene that determines pea shape. So the genetic constitutions—or GENOTYPES—of the pure-breeding round plants and wrinkled plants will be RR and rr respectively.

* All the offspring of this cross will have the genotype Rr, since they will all receive an R gene from one parent and an r gene from the other. Though the Rr plants carry both types of gene, they produce round peas, because the dominant "round" gene masks the effect of the recessive "wrinkled" gene.

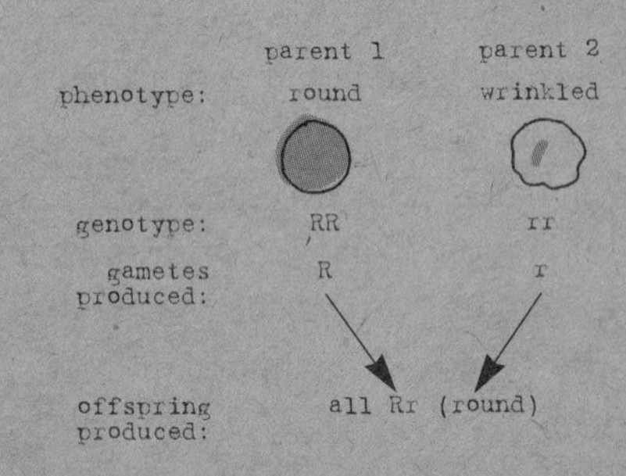

KEY WORDS

GENOTYPE:
the genetic make-up of an individual (e.g. RR, Rr, or rr), used with reference to a particular characteristic

PHENOTYPE:
the physical expression of an organism's genes (i.e. round or wrinkled)

HOMOZYGOUS:
possessing a pair of identical genes (i.e. RR and rr plants are homozygous)

HETEROZYGOUS:
possessing a pair of different genes (i.e. Rr)

ONE IN FOUR

* But what happens when two Rr plants are crossed with one another? The offspring can now inherit either an R or an r from each parent. R-bearing and r-bearing gametes are produced in equal numbers, so there is a 50 percent chance that an individual will inherit a particular gene from each parent. On average, you would expect one in four of the offspring to be RR, one in four to be rr, and two in four to be Rr. In other words, you would expect three round plants to be produced for every wrinkled plant.

gene from parent 2 ↓ / gene from parent 1 →	R	or	r
R	RR round		Rr round
or			
r	Rr round		rr wrinkled

A Matter of Chance

Inheritance is a game of chance, based on simple probabilities. Which of the two genes an individual receives from a parent is like flipping a coin and seeing if it comes up heads or tails. For example, the chance of inheriting an R gene from both parents is the same as the chance of tossing a coin twice and it coming up heads both times—i.e. one in four.

TOO GOOD TO BE TRUE?

* In the 1930s, an analysis of Mendel's results showed that statistically they were too good to be true. His figures were improbably exact. No one was suggesting that his results were faked, since his rules of inheritance were amply confirmed by other scientists.

* What is more likely is that he chose to "ignore" the results of any crosses that deviated too markedly from the expected 3:1 ratio.

take a chance?

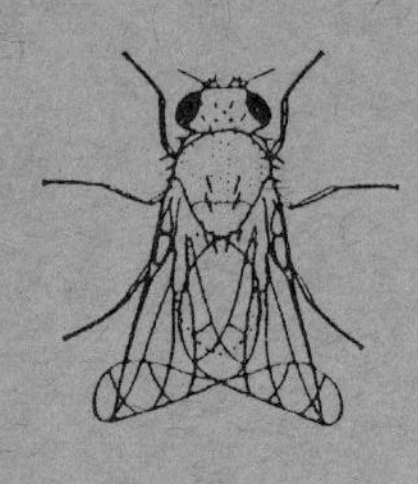

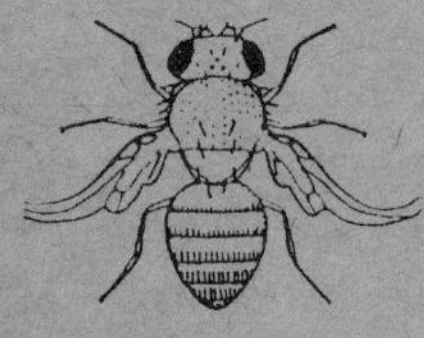

fruit flies

LINKING GENES TO CHROMOSOMES

* By the turn of the century it had been established that chromosomes came in pairs, just like Mendel's genes. This reinforced the growing belief that genes were actually carried on the chromosomes. But the final proof came from experiments on fruit flies by the American geneticist T.H. Morgan.

T.H. MORGAN

Thomas Hunt Morgan (1866-1945) came from a prominent family line. His great-grandfather F.S. Key composed the U.S. national anthem. When he began his experiments, he was initially doubtful of Mendel's laws—but the results of his own experiments, together with those of his students Alfred Sturtevant and Hermann Müller, convinced him that Mendel had been right all along.

THE FLY FACTOR

* **Morgan** had been studying the inheritance of fruit fly characteristics in much the same way as Mendel studied peas. But he had noticed something peculiar. ***The inheritance of certain characteristics seemed to be affected by the sex of the offspring***.

* In one experiment he used two strains of flies, one with red eyes and the other with white. When male white-eyed flies were mated with female red-eyed flies, all the offspring had red eyes. So red eyes were dominant and white eyes recessive. But when male red-eyed flies were mated with female white-eyed flies the result was different: both white-eyed and red-eyed flies were produced in roughly equal numbers. What's more, all the white-eyed flies were male.

* Morgan recognized that males always inherited the eye color of their pure-breeding mother. White-eyed mothers produced white-eyed males; red-eyed mothers produced red-eyed males.
* In fruit flies, as in humans, females have two X chromosomes while males have an X and a Y. Males inherit their X chromosome from their mother, so Morgan reasoned that the only way to explain his results was if the gene for eye color was physically attached or linked to the X chromosome. ***The Y chromosome seemed to carry no genes at all. In males any gene on the X chromosome, whether dominant or recessive, was always expressed.***

X-ray

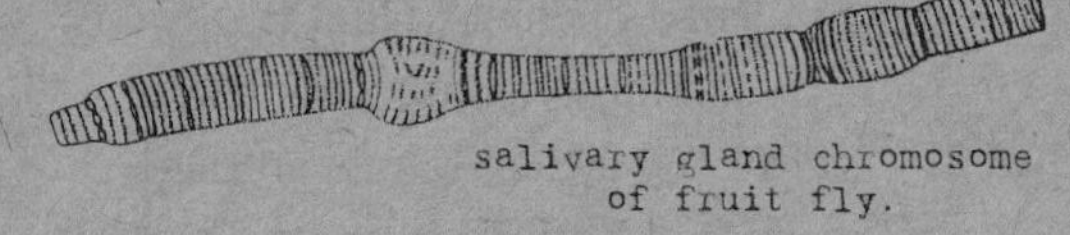

salivary gland chromosome of fruit fly.

WHY FRUIT FLIES?

* Measuring only a few millimeters in length, the fruit fly *Drosophila melanogaster* is, to look at, an unremarkable little creature. But it has become one of the workhorses of genetic research. Like peas, fruit flies produce lots of offspring—but they breed much more quickly.
* A fruit-fly egg becomes an egg-laying adult in about two weeks. In a year, it's possible to study over 20 successive generations of flies. Moreover, fruit flies have unusually large chromosomes in their salivary glands and there are only four pairs of them, so their inheritance is simpler to study.

Zapping the Fruit Flies

Hermann Müller *(1891–1966) discovered that X-rays could induce mutations in fruit flies. This enabled him to produce flies with lots of different visible characteristics and thus increase the scope of genetic studies.*

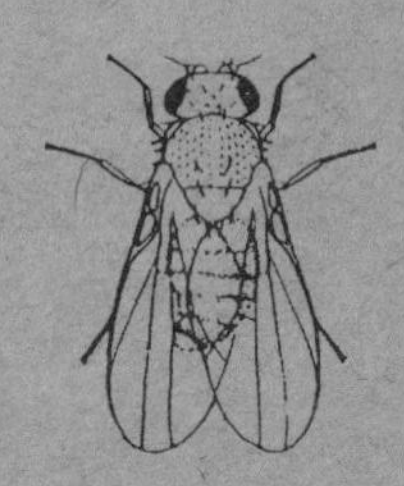

fruit flies are ideal for genetic research

BAD NEWS FOR MALES

* In humans the Y chromosome carries the male-sex gene, but little else. This is gloomy news for males. Any gene carried on the X chromosome will always express its effects, because it has no partner gene. So genetic diseases caused by X-linked genes are much more common in men than in women.

Royalty and Genealogy

Whatever your opinion of royal families, their contribution to human genetics has been invaluable. Because they have provided detailed family histories, in the form of royal pedigrees, they have enabled geneticists to trace the patterns of inheritance of many human characteristics and diseases.

ROYAL BLOOD

* HEMOPHILIA is often referred to as a royal disease because many of Queen Victoria's male descendants were afflicted with the disease—though the disease is not, of course, confined exclusively to royalty. The disease is characterized by an inability to make a vital blood-clotting factor, so that even the slightest cut results in severe bleeding. The defective gene responsible for the disease is recessive, but because it is carried on the X chromosome, males need inherit only one copy of the gene to get the disease, while women must inherit two copies. Not surprisingly, ***hemophilia is more prevalent in men than in women***.

Henry VIII

PEDIGREE CHUMS

* The hemophilia pedigree for Queen Victoria shows the incidence of hemophilia and its pattern of inheritance in some of her descendants. Women are represented as

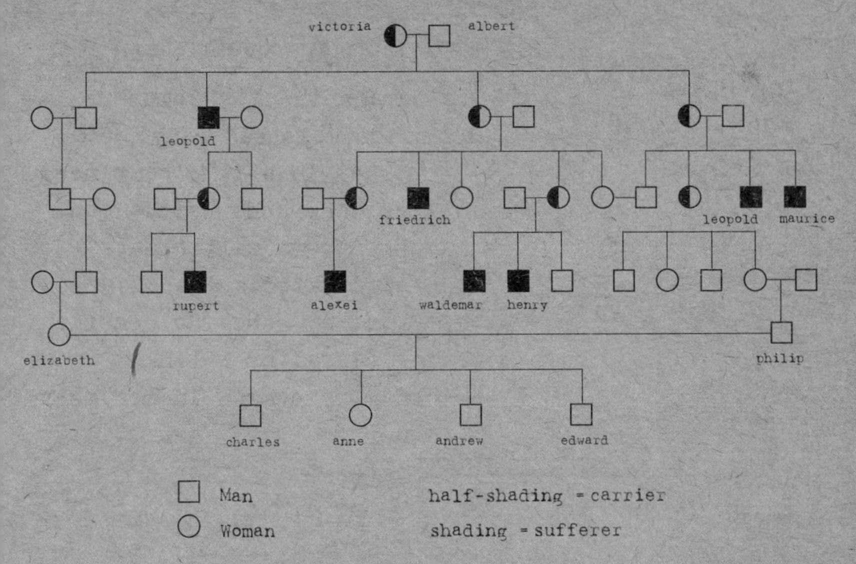

The hemophilia pedigree for **Queen Victoria**, showing the incidence of hemophilia and its pattern of inheritance in some of her descendants. It is obvious from the pedigree that the disease is X-linked, since all the affected individuals are males.

circles and men as squares. Individuals affected by the disease are shaded; carriers are half-shaded. Each separate row consists of members of the same generation. Horizontal lines connect parents; vertical lines indicate their descendants. It is obvious from the pedigree that the disease is X-linked, since all the affected individuals are males.

MEN ONLY

* Queen Victoria herself was unaffected by the disease, since she only carried a single copy of the gene, whose effects were masked by the normal dominant gene on her other X chromosome. The most famous sufferer was one of Victoria's great-grandsons, Tsarevitch Alexei of Russia.

you only want me for my DNA

Pioneer mapmaker

In 1911, Alfred Sturtevant (1891–1970), a student of Morgan's, reasoned that genes must be linearly arranged on the chromosomes, like a string of beads. The gene for any particular characteristic had a fixed location, or **locus**, on the chromosome. The first chromosome map that Sturtevant drew up for the fruit fly showed the position of five sex-linked genes. By 1922, he had produced a map showing 2,000 genes on the fruit fly's four pairs of chromosomes.

GENETIC ASSOCIATIONS

* Mendel had observed that genes for different characteristics, such as pea shape and flower color, behaved independently of one another. Whether a plant had round peas had no influence on whether it had white or pink flowers. But Morgan soon found exceptions to this rule.

EYES AND WINGS

* The explanation was simple. *If gene codings for different characteristics resided in different chromosomes, they behaved independently of one another—but if they were carried on the same chromosome, they tended to be inherited together.* When Morgan crossed flies possessing white eyes and short wings with flies that had red eyes and normal wings, he noticed that the white-eye and short-wing characteristics were inherited together.

* None of the offspring had white eyes and normal wings. Morgan reasoned that the genes for eye color and wing shape must reside on the same chromosome; and because they were physically linked with each other, they must be inherited together.

* By chance, Mendel had chosen to work on characteristics that were unlinked. Only when two genes were carried on different chromosomes did they behave independently of one another. But Morgan

soon found that linked genes did not always stay together. Every now and then, offspring with white eyes and normal wings would appear. ***The chromosomes within each pair seemed to be shuffling their genes with one another to create new genetic combinations.***

simple map of one of the *Drosophila* chromosomes

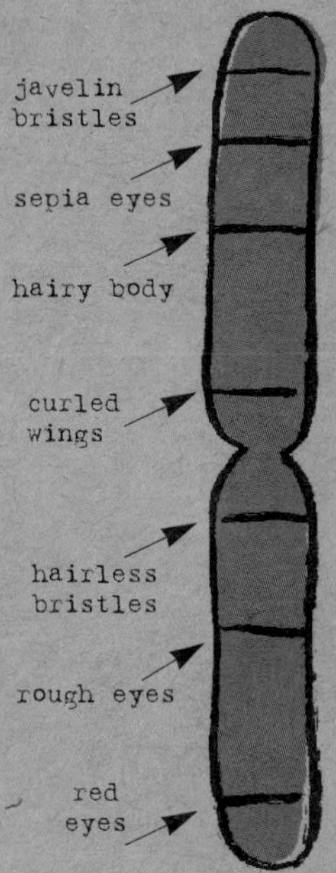

javelin bristles

sepia eyes

hairy body

curled wings

hairless bristles

rough eyes

red eyes

SHUFFLE THE DECK

* Morgan had discovered GENETIC RECOMBINATION, which occurs during the process of meiosis—when cells divide to form gametes. As a cell is preparing to divide, the homologous chromosomes line up alongside each other. They then pair up at various points along their length and exchange complementary segments of genetic material (see page 22 and 23). ***The chances of two linked genes being separated from each other during genetic recombination depended on their relative positions to each other on the chromosome.***

GENETIC MAPS

* Recombination, it seemed, was a bit like shuffling two packs of cards. Genes that were closer together on the chromosome were less likely to get separated. By looking at the frequency of new genetic combinations in the offspring of a cross, biologists could start to work out the relative positions of linked genes on the chromosomes and so begin making the first genetic maps.

COUNTING AND MEASURING

* In trying to get to grips with the basics of genetics, Mendel had deliberately chosen to work on simple characteristics that varied in an "either/or" fashion. Plants had either yellow seeds or green seeds, pink flowers or white flowers, for example. There were no intermediate types, and the offspring always resembled one of the two parents. But most characters vary in a more gradual and continuous fashion.

VARIATION

The difference between discrete and continuous variation is essentially the difference between counting and measuring. Analyzing the inheritance of discrete characteristics relies on counting the number of individuals of each distinct type, whereas analyzing the inheritance of continuous characteristics relies on measuring individuals.

height is a continuous characteristic

TALL OR SHORT

* In humans, most characteristics can't be divided up into a small number of DISCRETE (DISTINCTLY DEFINED OR SEPARATE) alternatives. Think of height, weight, and skin color, for example. These all display a much more continuous spectrum of variation.

* To the early geneticists, the inheritance of such "CONTINUOUS" CHARACTERISTICS seemed to have little to do with the simple rules laid out by Mendel. Tallish fathers and shortish mothers invariably produced children of intermediate height. It therefore appeared, superficially at least,

that inheritance was a process of blending. But the blending idea could not explain how parents of a medium height were able to produce children who were tall or short.

offspring may inherit a balance of parental characteristics

MANY GENES

★ ***Nowadays, biologists recognize that "continuous" characteristics are controlled by genes at many different loci.*** Though they may sometimes give the impression of blending, each one of the gene pairs behaves in a similar way to the gene pair responsible for controlling pea shape. Skin color, for example, is influenced by at least four gene pairs. Each gene in a gene pair makes a contribution to the overall expression of the skin color, either in the lighter or darker direction—so that the skin color you inherit will be the sum of the effects of several separate genes.

The Bell Curve

If you took a random sample of people, measured all their heights, and then plotted the number of people at each height, the resulting frequency distribution would resemble a bell-shaped curve called a normal distribution. The population would display a continuous range of heights, though most people would cluster around the average. Most human characteristics display a similar pattern of variation.

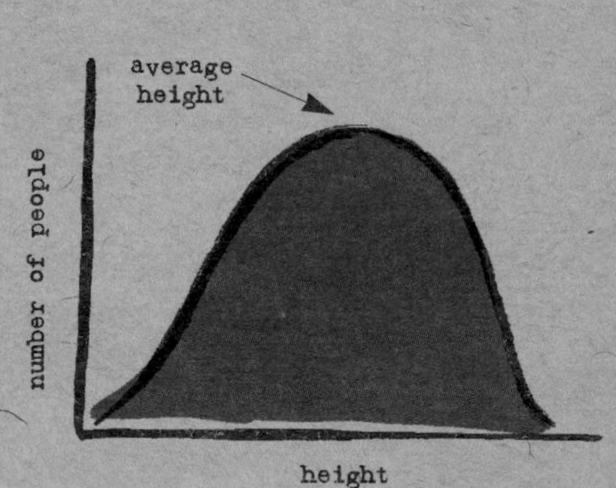

frequency distribution of heights in the human population

uneducated workers were believed to have a low IQ

THE STAIN OF EUGENICS

* The study of human inheritance was established by Francis Galton. But Galton's interest had as much to do with social prejudice as it did with hard science. Galton was obsessed with the biological improvement of the human race. For many of Galton's supporters, the rediscovery of Mendel's laws seemed to give scientific credibility to their racist and bigoted views. Genetics was about to enter the bleakest period of its short history.

IQ Tests

The IQ (Intelligence Quotient) test was initially established as a way of discriminating between "able" and "unable" people. Because intelligence was believed to be an innate genetic quality, early versions of the test conveniently overlooked the influence of education. This had the effect of highlighting the apparently low intelligence of the poorer classes, for whom quality education was a pipe dream.

SLUMMING IT

* In the 19th century, divisions between rich and poor had been exacerbated by the mass urbanization that accompanied the industrial revolution. Slums had become endemic to many of the world's largest cities. To those with wealth and status, poverty was seen as an expression of lesser ability. Slums were considered breeding grounds for the worst characteristics humanity had to offer.

* After conducting a survey of prominent and wealthy families, Galton claimed in his book *Hereditary Genius* (1869) that ability was inherited—able fathers had able sons—and could not be enhanced by education. Galton was the first to suggest that the proliferation of the poor and underprivileged was getting out of hand. He believed that the

the inferior race threatens the future of society—the state must intervene

uncontrolled spread of an "inferior" race threatened the future of society, and that it was time for the state to intervene.

* In 1883, he coined the term "EUGENICS" to describe the practice of biologically improving the human race. There were two ways in which he envisioned this could be done. First, positive eugenics would encourage the professional classes to have more children by granting income tax relief for each child born. Second, and more controversial, negative eugenics aimed to physically prevent those of lesser ability from having children.

SCIENCE GONE MAD

* Initially Galton's proposals received a cool reception. But for many who shared his influence and wealth the reemergence of Mendelism added scientific weight to his arguments. Criminality, mental illness, and low intelligence were seen as products of specific genes—and only by preventing inferior races from breeding would these genes be removed from circulation.

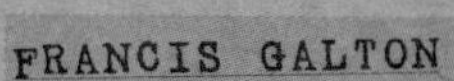

FRANCIS GALTON

Francis Galton (1822–1911) founded the Laboratory for National Eugenics (now the Galton Laboratory) at University College London, the first human genetics department in the world. His often eccentric career also included the scientific study of fingerprints, statistical tests on the efficacy of prayer, and the publication of a human-beauty map of the British Isles.

any disability was a mark of "inferiority"

BLAZING A TRAIL OF BIGOTRY

* The first eugenics legislation was introduced in the early 20th century, and concentrated mainly on negative eugenics. Its ramifications, which were felt throughout Europe and the U.S., culminated violently and abruptly in the comprehensive sterilization programs of Nazi Germany.

INFERIORITY COMPLEX

* Concerned that uncontrolled immigration of "racially inferior types" was threatening the genetic health of Americans, many states imposed quotas on the number of immigrants allowed to settle. It is a terrible irony that by the 1930s, Eastern Europeans— one of the primary targets of US prejudice— were seeking new homes in order to escape from the even more extreme racist program orchestrated by the Nazis.

PHONEY SCIENCE

* Adolf Hitler's obsession with racial purity and the Aryan ideal took the prejudice and bigotry of eugenics to a grim conclusion. Anyone with any form of mental or physical handicap was

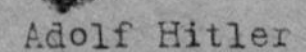

Adolf Hitler

Sterilization of inferior types

Many U.S. states passed legislation requiring the incarceration and/or sterilization of so-called inferior types. This was an umbrella term that included the mentally ill or insane, people of low intelligence, and criminals. But prejudice was given a free reign, and in some states the inferior label was extended to include homosexuals and Communists! By the 1930s, at least 20,000 people had been sterilized in the U.S.

institutionalized, man, woman, or child. And in Nazi Germany incarceration usually meant extermination, rather than sterilization. Hitler may have occasionally dressed his policies in the language of science, but the "science" of eugenics was now no more than a sham.

eugenics was an excuse to persecute people of low intelligence

"America must be kept American. Biological laws show that Nordics deteriorate when mixed with other races."

CALVIN COOLIDGE, PRESIDENT OF THE UNITED STATES 1923-29

DIRTY POLITICS

* To assume that all human traits were exclusively controlled by specific genes that could be artificially selected was not only deeply simplistic, but also untrue. Today, eugenics remains a blemish on the history of genetics. It was a political movement, with little or no grounding in science. The study of human genetics fell into a long period of decline because scientists became reluctant to associate themselves with a subject tarnished by controversy and human suffering.

A ONE-WAY TICKET

* Although hereditary determinism was all the rage in the West, in the Soviet Union of the 1930s its political antithesis—Lamarckism—was the dominant force. This time it was biologists who bore the brunt of political ideology. Mendelism was perceived as a bourgeois capitalist conspiracy against Marxism, and any Soviet scientists sympathetic to Mendelian genetics were given a one-way ticket to Siberia.

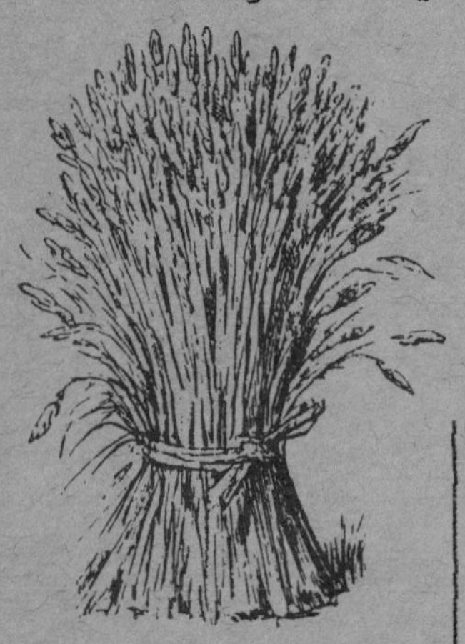

wheat was a vital crop in the Soviet Union

MARXIST PHILOSOPHY

* Stalin despised the idea that even pea shape, let alone any human characteristic, was determined by genes. Lamarckian inheritance sat much more comfortably alongside the Marxist philosophy, which emphasized the importance of the environment in shaping the individual and society.

Mendelian Martyr

In 1940 **Nikolai Vavilov** *(1887–1943), one of the Soviet Union's best Mendelian geneticists, was arrested and interrogated for 1,700 hours. In a five-minute trial, he was found guilty of crimes against the state. He died in a prison camp in 1943.*

★ Stalin had found a close political ally in the Russian biologist **Trofim Lysenko**. Lysenko had made a name for himself through his experiments on the "vernalization" of wheat—a process in which seeds are frozen so that they will germinate earlier the following spring. Although vernalization was already well known in the West, Lysenko claimed that the effect was inherited, in a Lamarckian manner. If true, this would have had immense agricultural value in areas where the growing seasons were particularly short.

"We shall go to the pyre, we shall burn but we shall not renounce our convictions."
NIKOLAI VAVILOV

Lysenko claimed he could protect plants from cold weather

SHORT, SHARP SHOCK

★ Lysenko pledged to offer scientific short-cuts to the perennial problem of food shortages. By exposing crops to cold shocks, for example, he claimed that new and hardy ice-resistant crops would evolve.

snowflakes

As the newly instated Director of Agriculture, Lysenko gained sufficient political and scientific influence to impose Lamarckian inheritance wholesale on Russian biology. Geneticists were given a stark choice: either deny their allegiance to Mendelism and its capitalist subplot or face imprisonment in a Siberian labor camp.

Lysenko's promise

Lysenko's promise to revolutionize Soviet agriculture eventually came to nothing. But not before he had managed to purge Russia of its best geneticists. His ideas, rooted in political ideology rather than science, effectively cut the Soviet Union off from the scientific advances being made in the West. Only in the 1950s did Soviet genetics begin to return to normality.

FROM YELLOW TO BLACK, FROM GENES TO ENZYMES

* The first hint that genes and enzymes—protein catalysts that increase the rate of chemical reactions in the body—were related to each other dates back to a breakthrough made by the English physician Archibald Garrod in 1908. Garrod discovered that an inherited disease was caused by a block in a chemical reaction within the body and hypothesized that a functioning gene produces a specific enzyme.

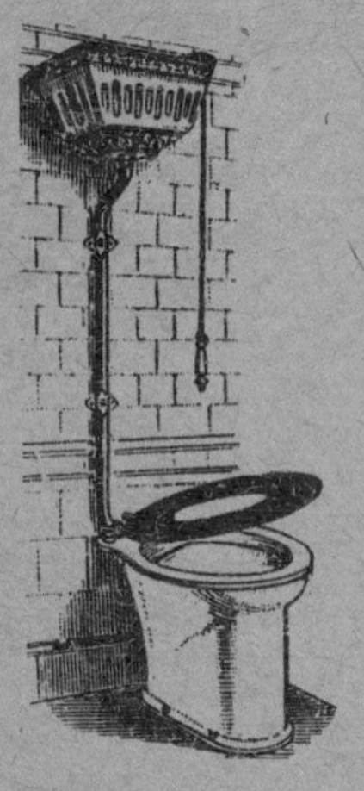

the evidence is in the can

BLACK STUFF

* Garrod had been fascinated by the rare genetic disease ALKAPTONURIA, whose inheritance followed a simple Mendelian pattern. Though the disease is not serious, its symptoms are striking. After eating certain types of food, affected persons

Garrod in his laboratory

Essential Acids

Not all the amino acids needed to make proteins can be synthesized by the body itself. About eight of the 20 or so amino acids that the body needs must be provided in the diet. These are known as essential amino acids.

produce urine that turns black as soon as it is exposed to the air. The black color is due to the presence of homogentisic acid, a chemical intermediate produced in the metabolic breakdown of the amino acid tyrosine. In persons unaffected by the disease, homogentisic acid is further broken down, and eventually converted to carbon dioxide and water.

essential amino acids are provided in diet

ENZYME ABSENCE

* Garrod suggested that alkaptonuria was caused by the absence of a specific enzyme. The missing enzyme would mean that the normal sequence of chemical reactions in the breakdown of tyrosine would be blocked, resulting in the accumulation of homogentisic acid in the blood, tissues, and urine. It was later confirmed that people affected by alkaptonuria do in fact lack the enzyme needed to break down homogentisic acid.

* ***Garrod had shown that a mutation in a particular gene produced a defect in a particular enzyme—which led, not unnaturally, to his conclusion that a functioning gene specifies an enzyme that enables a chemical reaction to take place in the body.***

* His work was later confirmed by the American scientists George Beadle and E.L. Tatum in the early 1940s.

NEGLECTED BREAKTHROUGH

Archibald Garrod (1857–1936) was the first to show a connection between an altered gene and an enzyme—one of the fundamentals of genetics. His work, however, like Mendel's, went pretty much unnoticed for almost 30 years.

CHAPTER 2

THE DNA BREAKTHROUGH

* If genes made proteins, what were genes made of? DNA was not an obvious candidate. Most biologists believed that it was chemically too simple to encode genetic information. Proteins themselves, with their long chains of 20 different kinds of amino acids, were considered the only molecules complex enough to be the genetic material. But similar studies by two different scientists suggested otherwise. DNA was the genetic material, in bacteria at least.

KEY WORDS

PATHOGEN: any organism that causes a disease

VIRULENCE: an organism's disease-causing ability

TRANSFORMATION: a change in an organism's characteristics due to the incorporation of "foreign" DNA into its own genetic recipe

SMOOTH OPERATORS

* In 1928, the English scientist **Frederick Griffith** was working at the Ministry of Health in London, investigating new techniques for classifying PATHOGENS. Griffith had become particularly interested in the PNEUMOCOCCUS bacteria, which cause pneumonia. The bacteria came in two distinct strains, a "smooth" strain and a "rough" strain. Griffith noticed that only the smooth strain caused the disease. When mice were injected with the smooth strain, they soon dropped dead, while injecting the rough strain had no effect on the health of the mice.

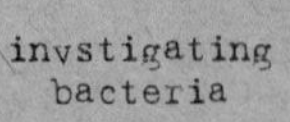

invstigating bacteria

* Griffith then discovered that heating the bacteria was a sure way of killing them. So when heat-killed smooth-strain bacteria were injected, the mice remained fit and healthy. But, amazingly, when he injected a mixture of heat-killed smooth-strain, and living rough-strain bacteria, the mice contracted pneumonia and died. Inside these dead mice, he found living, smooth-strain bacteria. Griffith reasoned that a chemical substance in the heat-killed bacteria must have transformed the rough strain into the virulent form.

* Intrigued by Griffith's work, in 1944 **Oswald Avery**, working at the Rockefeller Institute in New York, refined Griffith's experiments and identified the "TRANSFORMING PRINCIPLE" as DNA. Even this apparently conclusive evidence was not enough for some scientists, who claimed that the bacterial transformation must have been caused by protein contaminants in the transferred DNA.

FREE-FLOATING DNA

When bacteria are killed by heat, their cell walls break down and their DNA is released into the environment. Living bacteria can change their characteristics by taking up this free-floating DNA and incorporating it into their own genetic recipes.

PERSUASIVE EVIDENCE

By the early 1950s, other bits of evidence had accumulated that seemed to support the role of DNA as the genetic material. It was discovered, for example, that the DNA content of sperm and eggs was half that of the somatic cells. Since it was already accepted that gametes contained half the normal number of chromosomes and that genes were carried on chromosomes, the argument that DNA was the stuff of which genes were made looked persuasive.

MORE CLUES ABOUT DNA

★ In the years that immediately followed Avery's discovery, all sorts of circumstantial evidence turned up that seemed to confirm that DNA was the genetic material in other organisms. But the strongest evidence so far came from the study of viruses. In 1952, Alfred Hershey **and** Martha Chase **were conducting some experiments on bacteriophages (or "phages," for short)—viruses that infect bacteria.**

SIMPLE SOULS

★ "Complex" is not a word commonly used to describe viruses. They are simplicity itself. ***A virus amounts to nothing more than a few genes encased in a protein shell.*** In fact, some scientists doubt whether viruses qualify for the term "living." Unlike true biological cells, they do not multiply by dividing in two. Instead, they rely on the cellular machinery of other organisms to reproduce.

★ By marking the protein and DNA constituents of a phage with specific and distinct radioactive labels, Hershey and

caution:
virus at work

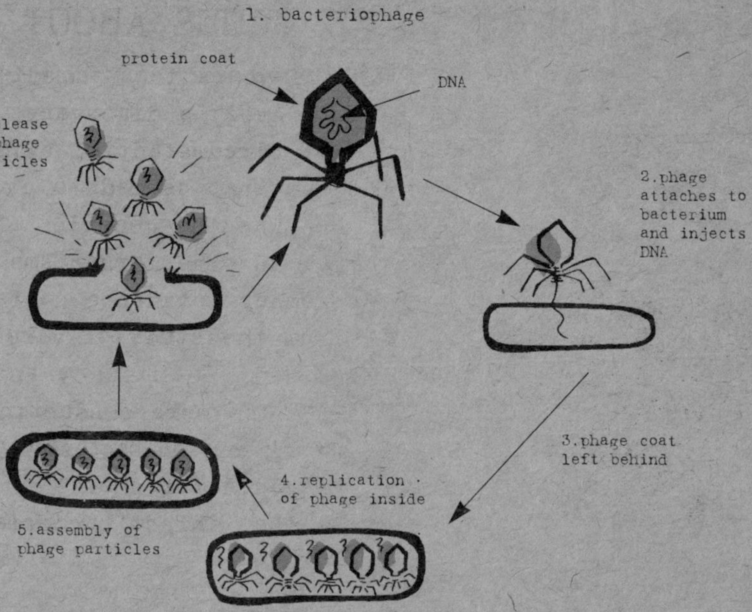

the short and sordid life of a bacteriophage

Chase were able to show that only the DNA is necessary for the reproduction of the phage.

* When a phage latches on to the wall of a bacterial cell, it hypodermically injects its DNA into the bacterium. The protein coat gets left behind. After a while, the bacterial wall is ruptured, releasing hundreds of newly formed virus particles.

* Hershey and Chase showed conclusively that the ***phage's DNA contains all the information needed to make a new virus particle— including its DNA and its protein coat***.

Viruses

Most viruses carry about 50 genes inside their protein coats, though some have as few as three and others as many as 300. Viruses are responsible for some fierce human diseases—including measles, mumps, chickenpox, polio, smallpox, and AIDS.

DECEIVED BY DNA

★ Skepticism about DNA's role in heredity was not as mind-bogglingly dim as it might seem today. Some scientists simply could not envision a way in which a molecule comprising a seemingly monotonous linear arrangement of four basic building blocks could encode the vast amount of information needed to produce a complex living thing. Much of the confusion stemmed from a fundamental misunderstanding of DNA's chemical make-up.

RNA AND DNA

A nucleic acid is a complex organic compound found in living cells, made up of a long chain of nucleotides. There are two types of nucleic acid–DNA and RNA. In RNA the nucleotides consist of a sugar molecule (subtly different to the one in DNA) and a phosphate molecule bound to one of four bases: Uracil, Cytosine, Guanine, and Adenine. So three of the RNA bases are identical to those in DNA.

BORING OR BOGUS

★ By the 1940s, the Russian-born chemist **Phoebus Levene** had shown that there were *two kinds of nucleic acid—DNA and its closely related sister molecule* RNA (RIBONUCLEIC ACID)*—and had defined the chemical differences between them*.

★ DNA was made up of long chains of nucleotides. But crude chemical analyses of DNA had led several scientists to the erroneous conclusion that DNA was composed of exactly equal amounts of each of the four nucleotides. The "tetranucleotide hypothesis" was originated by the German chemist **Albrecht Kossel**, but is nowadays most closely associated with Phoebus

Levene. At the time, the tetranucleotide hypothesis helped to create the popular image of DNA as a regular and uninteresting molecule, with little relevance to genetics.

all animals and plants have RNA and DNA

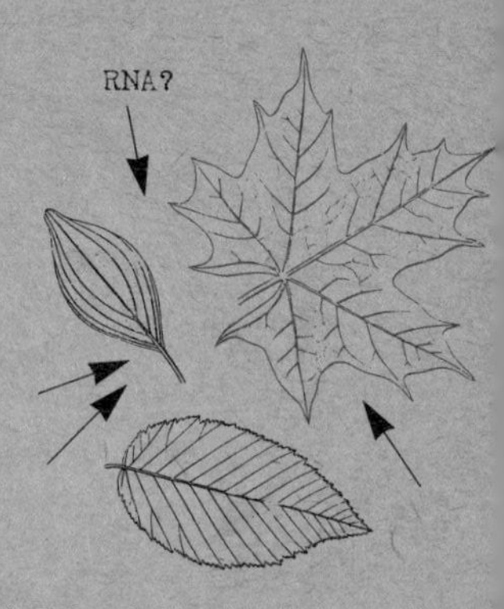

A simple mistake

For many years, scientists mistakenly believed that DNA was found only in animals and RNA was found only in plants. When RNA suddenly turned up in animal cells, scientists initially suggested that it must have come from the animal ingesting a plant. Today, everyone agrees that plants and animals have both DNA and RNA.

PERSUADED BY PROTEINS

* Levene was convinced that proteins held the key to the genetic code. He was a contemporary of Avery's at the Rockefeller Institute in New York, and also one of his fiercest critics. Levene's acceptance and dogmatic adherence to the tetranucleotide hypothesis has been heavily maligned. Some have even accused him of delaying the progress of genetics research by 20 years or more.

X-RAY SPECS

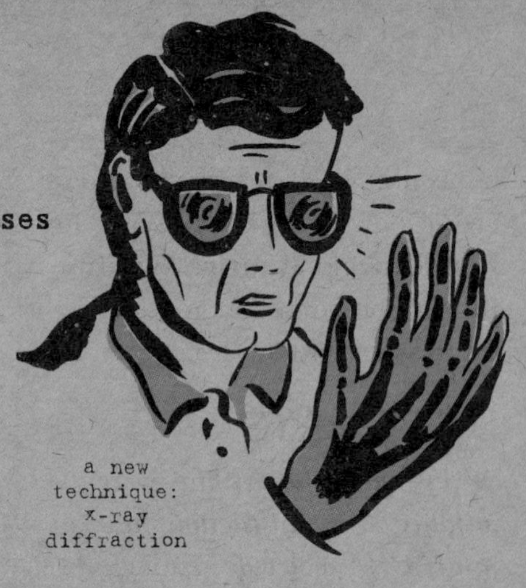

a new technique: X-ray diffraction

★ Crude chemical analyses of DNA had involved breaking up the DNA molecule into its constituent parts, thereby destroying its most interesting aspect—its shape. To understand the way in which DNA worked, scientists needed to obtain a detailed picture of an intact DNA molecule. An exciting new technique known as X-ray diffraction would soon make this possible.

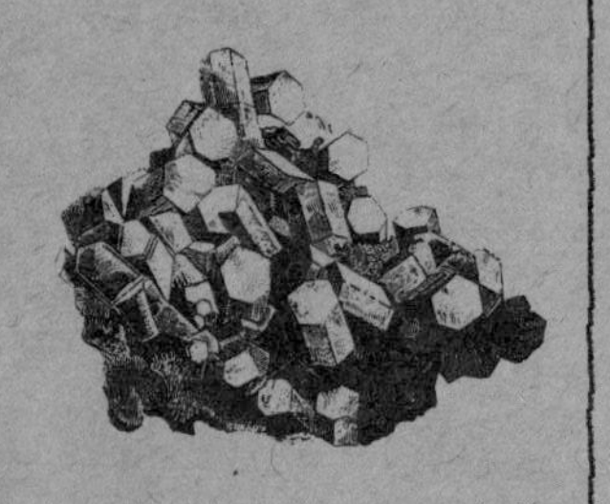

James Watson and Francis Crick were to put the "finishing touches" to the DNA jigsaw, and take all the glory.

CRYSTAL CLEAR

★ X-ray diffraction had originally been developed by physicists to probe the fine internal structure of crystalline substances. It involved firing a beam of X-rays at a crystal and recording the pattern of reflected rays on an X-ray film.

★ Using some fancy mathematics, the specific pattern of reflected rays could be used to interpret the precise arrangement and position of the atoms within the crystal.

★ The technique worked best for molecules with extremely regular arrangements of atoms, so crystals were ideal. But some believed that it might also be possible to use X-ray diffraction to study the structure of more complex biological molecules, such as proteins and maybe even DNA. By the early 1950s, **Rosalind Franklin** and **Maurice Wilkins**, both at King's College in London, had set to work on this very task.

DNA X-RAYS

★ ***Franklin and Wilkins were the first to produce X-ray diffraction patterns of an intact DNA molecule***. Though the patterns in the X-ray photos were a little vague and fuzzy, they did at least hint at some degree of regularity in the structure of the DNA molecule. Franklin suggested that the most likely explanation for the patterns she observed was if DNA was a helix-shaped molecule made up of 2, 3, or 4 closely compacted nucleotide chains.

CAMBRIDGE CELEBRITIES

★ Although Franklin had gotten close to the truth, from now on she would take a back seat in the DNA story. Her name was soon to be eclipsed by the celebrity of two scientists who had joined another X-ray diffraction group at the Cavendish Laboratory in Cambridge. **James Watson** and **Francis Crick** were to put the "finishing touches" to the DNA jigsaw, and take all the glory. Such is often the way in science, and few people have heard of Rosalind Franklin.

ROSALIND FRANKLIN AND MAURICE WILKINS

The relationship between **Rosalind Franklin** (1920–58) and **Maurice Wilkins** at King's College was strained—perhaps because their differing responsibilities in the DNA work were so ill-defined. Franklin didn't live to witness the unfolding DNA story. She died from cancer at the age of 37, and was excluded from the Nobel prize shared by Wilkins, Crick, and Watson in 1962.

the DNA jigsaw was nearly complete

HOMING IN ON THE HELIX

* James Watson and Francis Crick first met in 1951. Crick was a 35-year-old physicist who had joined the Cavendish Laboratory to study X-ray diffraction theory. Watson, an American biologist more than 10 years his junior, was convinced of DNA's importance in heredity, though he had little knowledge of chemistry or crystallography.

HELP FROM PAULING

The publication of a scientific paper on helical molecules, written by the famous American chemist **Linus Pauling**, had an important influence on Crick. He promptly immersed himself in some complicated mathematical theory, which he believed might help him interpret the diffraction patterns from DNA.

THREE DIMENSIONAL

* The two men thrived on one another's company and set to work on interpreting Franklin's DNA diffraction patterns. Interpreting the three-dimensional structure of a molecule from a few scattered dots on a piece of X-ray film was not easy. In fact, it was a fine mathematical art. As far as DNA was concerned, Franklin had already narrowed down the possibilities, but a helical molecule can have any number of different pitches and diameters. And was it one, two, three, or four helical chains of nucleotides?

* The two men spent much of their time together building prototype models of DNA molecules out of bits of brass, wire, and screws. The idea was simple, in theory at least. For each model, they would make predictions, using Crick's newly learned

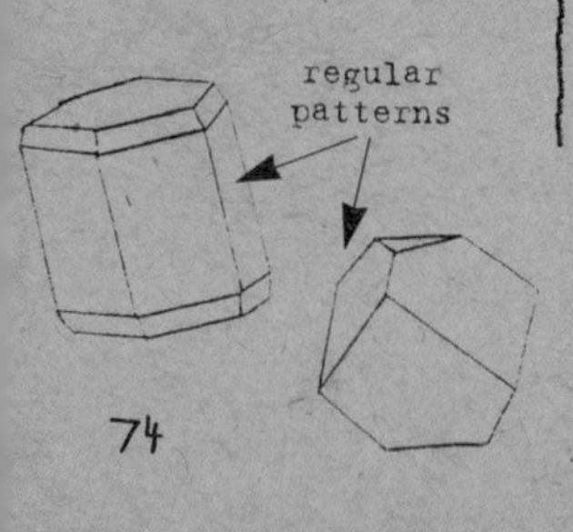

mathematical skills, as to what the expected diffraction pattern would be. And they would continue building different helical models until they came up with an expected pattern that exactly matched the one observed with a real DNA molecule.

Watson and Crick building models

* As they worked, they were constantly acquiring new facts about DNA from other sources. One of the most important pieces of information came from the Austrian chemist **Erwin Chargaff**, then working at Columbia University in New York, with whom they shared a mutual antipathy. ***Chargaff had observed that there was something strikingly regular about the chemical composition of DNA. The number of A's always matched the number of T's, and the number of G's always matched the number of C's. This was a pretty big clue that A and T, like G and C, might form complementary pairs within the DNA molecule.***

Easy when you know how

"X-ray diffraction is not a difficult branch of physics: on the contrary, it is easy to the point of tediousness. The widespread view that it is unintelligible has arisen because a certain intellectual effort is needed to grasp its mathematical foundations, and because it is supposed, incorrectly, that some special type of 'three-dimensional imagination' is a prerequisite for understanding its methods and results."
Francis Crick

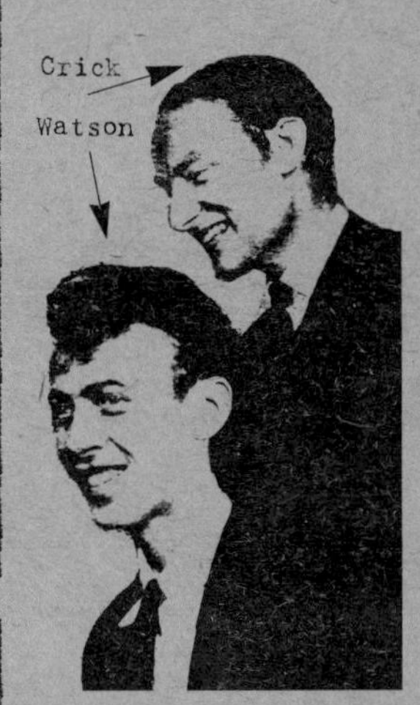

Crick and Watson were very pleased with themselves

DNA EXPOSED

* One day in 1953, Watson and Crick experienced the eureka moment. They came up with a structure for DNA whose predicted X-ray diffraction pattern matched the one from the real molecule perfectly. The DNA double helix was announced to the world.

"It has not escaped our notice that the specific pairing we have postulated immediately suggests a possible copying mechanism for the genetic material."

From an article by Watson and Crick in *Nature*, 1953

SIMPLE STUFF

* The structure that Watson and Crick ended up with looked complicated on first acquaintance. But if you ignored the insignificant details, you were left with a model of beautiful simplicity.
To visualize its structure, imagine (or make) a straight ladder, about 6 inches in length and made out of plasticine. If you hold the ends and rotate them in opposite directions, you will end up with something roughly resembling a spiral staircase (minus the handrail!).
* The DNA molecule consists of two intertwining strands, held together by the complementary base pairings. Think back to the straight version of the plasticine ladder again. If you were to cut the ladder in half down the middle of each rung, then each half of the ladder would represent one of the (uncoiled) DNA strands. Each

strand is made up of an unbroken chain of nucleotides. Alternating SUGAR and PHOSPHATE molecules form the backbone of the strand (the vertical side pieces of the ladder), while the BASES—the DNA letters—project inward (the half-rungs of the ladder). In the DNA molecule itself, the two strands are held together by the CHEMICAL BONDS that form between the complementary letters.

the helical DNA molecule

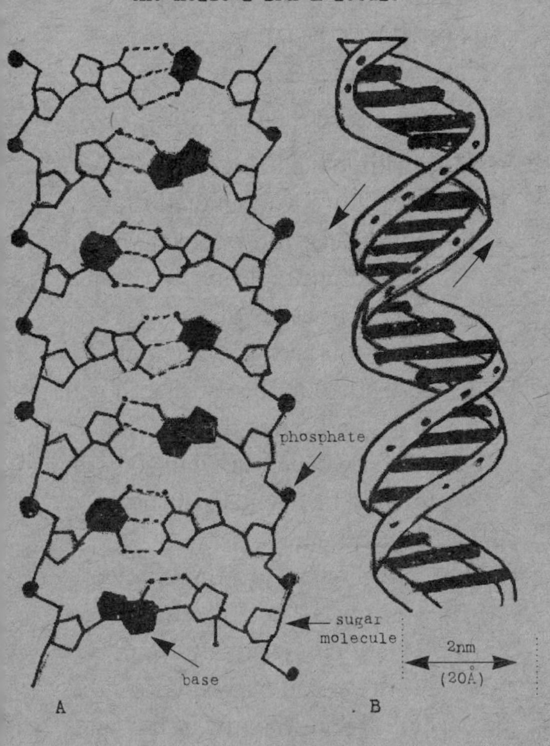

Diagram A shows the helical DNA molecule opened out with the base pairings in the middle. Diagram B shows the spiral staircase shape of the DNA molecule, with the base pairings as the rungs of the ladder.

Underwhelmed

Not everyone was excited and overwhelmed by the discovery of DNA's structure. Chargaff, who had never seen eye to eye with the dynamic duo, was positively underwhelmed. Summing up his feelings, he remarked with consummate disdain: "That in our day such pygmies throw such giant shadows only shows how late in the day it has become."

STRUCTURAL INSIGHT

The structure of DNA immediately hinted at a way in which the molecule could replicate itself. Because of the complementary base pairings, the base sequence on one strand automatically determined what the base sequence on the other strand must be. So each single strand could act as a template for the replication of a new strand.

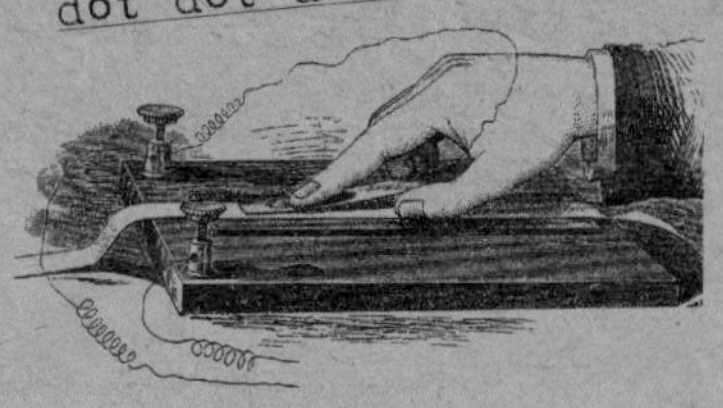

DECIPHERING THE CODE

★ Watson and Crick had removed any lingering doubts that DNA was the genetic material. But there was still an important question that had to be answered. How did the DNA encode genetic information? Not content with one revolutionary discovery, Watson and Crick turned their attention to the next big question in genetics.

INDUCED MUTATIONS

★ Watson and Crick were joined by another Cambridge scientist, **Sydney Brenner**, and all three began studying the effect of induced mutations on the growth of phages. By using chemicals or radiation they could aim mutations at specific points in the DNA of the phages' genes. They could then look at what effect these mutations had on the coded products—i.e. the proteins—by observing how well the phages grew in the laboratory.

★ They soon discovered that ***the phages' growth depended on how many bases were affected by the mutation***. Inserting three extra bases in a gene had little effect on growth— but if only one or two bases were changed, then the phages couldn't grow.

Breaking the DNA code

They suggested that the sequence of letters in a gene is read in order, from one end to the other, with each triplet of letters coding for an amino acid.

DNA sequence	AAT	CCG	GTA	TTG	ATA	CTC	CCC	ATT
amino acids	aa1	aa2	aa3	aa4	aa5	aa6	aa7	aa8

Inserting an extra triplet of letters into a gene would cause an extra amino acid to be inserted into the amino-acid sequence. Such a minor change to the protein's structure would have little effect on the phages' ability to grow properly.

AAT	CCG	GTA	TTG	GAG	ATA	CTC	CCC	ATT
aa1	aa2	aa3	aa4	aa9	aa5	aa6	aa7	aa8

insertion of an extra triplet of letters causes an extra amino acid to be inserted; rest of amino-acid sequence remains unchanged

But the removal or insertion of either one or two bases would change every triplet sequence beyond the point where the mutation occurred. This would lead to drastic alterations to the protein's amino-acid sequence, reducing its ability to function and the phage would be unable to grow.

insertion of two bases changes entire amino-acid sequence beyond the insertion point

AAT	CCG	GTA	TCT	TGA	TAC	TCC	CCA	TT
aa1	aa2	aa3	aa10	aa11	aa12	aa13	aa14	

Crick and his coworkers had shown that the DNA code was based on triplets of letters. A DNA sequence was read one triplet at a time, from a fixed starting point.

RNA MOVES INTO PLACE

DNA was beginning to make sense

★ DNA was beginning to make sense. A gene was a sequence of three-letter words, each word coding for an amino acid. The amino acids were joined together in a long chain to form a protein. But there were still some pretty big gaps in the DNA story. How, for example, was the DNA code physically translated into an amino-acid sequence? It was time for DNA's sister molecule, RNA, to move into the spotlight.

LINES OF COMMUNICATION

★ Protein synthesis presented an intriguing puzzle. **Paul Zemecnik** and his colleagues from Massachusetts General Hospital in Boston had *discovered that proteins were manufactured in the* RIBOSOMES *(the cell's protein factories), which were situated outside the nucleus (the cell's control center). But the DNA, which encoded the information to make a protein, was located inside the nucleus.*

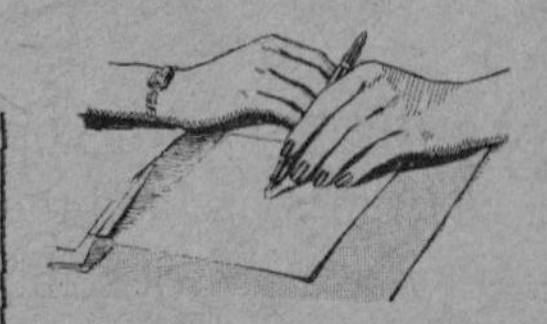

WHAT'S IN A LETTER?

In RNA, the base thymine (T) is replaced by uracil (U). But, like T, U forms a complementary pairing with adenine (A). U can be thought of as an equivalent version of T. So the four letters in RNA are A, G, C, U. Although the languages of RNA and DNA are subtly different, the coded information is identical.

What's more, there was no DNA in the ribosomes. So how did the genetic message get from the control center to the protein factory?

* RNA provided the key. RNA was a nucleic acid, like DNA, and it came in several different forms. MESSENGER RNA

lines of communication

(mRNA) made a copy of the original DNA message and transported it from the nucleus to the assembly line in the protein factories. Amino acids were then brought into the factory by TRANSFER RNA MOLECULES (tRNA) and bolted together in the correct sequence.

* So protein production came in two distinct phases. During the first phase (TRANSCRIPTION) the DNA helix unwound, and mRNA molecules floating around in the nucleus were attached to their complementary partners on one of the DNA strands. In the second phase (TRANSLATION) the transcribed message was "read" by tRNA molecules and converted into an ordered sequence of amino acids.

KEY WORDS

TRANSCRIPTION: the process by which a DNA message is converted into its equivalent mRNA

TRANSLATION: the process by which the sequence of bases in the mRNA is translated into a sequence of amino acids

RIBOSOMES: the cell's protein factories, situated outside the nucleus

amino acids are bolted into the correct sequence

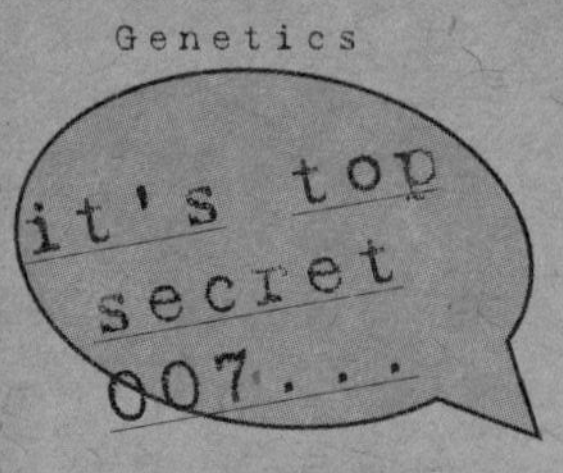

CODE BREAKERS

* Crick had identified a slight problem with his genetic-code hypothesis. If DNA was a language of three-letter words, and each word was made up of any of the four letters, then in total there were 64 (4x4x4) different possible words. Yet there were only 20 different amino acids. The number of words and amino acids didn't match up.

Puzzle solved!

The genetic code had been cracked and was shown to be universal. The same triplet of letters coded for the same amino acid, whether it came from the genetic recipe of a bacterium or a human.

SYNTHETIC SEQUENCES

* Enter stage right, **Severo Ochoa** from New York University medical school; and stage left, **Marshall Nirenberg** and his student **Phil Leder** from the National Institutes of Health in Maryland. By the early 1960s techniques in molecular biology were advancing rapidly, and it was now possible to design and make synthetic DNA molecules of any sequence that took your fancy.
* Using a cocktail of DNA, RNA, amino acids, and RNA POLYMERASE, it was even possible to manufacture proteins in a test tube. By changing the DNA code and then looking at the amino sequence of the manufactured protein, they could work out the relationship between each of the 64 possible triplets and the 20 amino acids.
* They discovered that although each amino acid was coded for by a different code

word, some of the 64 possible words coded for the same amino acid. In other words, the DNA code was DEGENERATE. If an amino acid had several different code words, the differences between the words was always found at the third letter in the triplet. Two of the code words also corresponded to start and stop signs, as Crick had predicted, which defined where in the DNA sequence a gene began and ended.

UUA UUG CUU CUC CUA CUG	UCU UCC UCA UCG AGU AGC	UUU UUC	UAU UAU
leucine	serine	phenyl-alanine	tyro-sine
UGU UGC	UGG	CCU CCC CCA CCG	CAU CAC
cysteine	trypto-phan	proline	histi-dine
CAA CAG GAA GAG	CGU CGC CGA CGG AGA AGG	AUU AUC AUA	ACU ACC ACA ACG
gluta-mine	arginine	isoleucine	threo-nine
AAU ACC	AAA AAG	GUU GUC GUA GUG	GCU GCC GCA GCG
aspara-gine	lysine	valine	alanine
GAU GAC	GGU GGC GGA GGG	AUG	UAA UAG UGA
aspartic acid	glycine	start!	stop!

LOOSE ENDS

Ochoa and Leder's observations tie in quite neatly with what is now known about mutations. Since the code was degenerate, a mutational change at the third letter of a word would not necessarily alter the amino-acid sequence of a protein. And sure enough, **most of the mutations that we observe in living organisms are found at the third letter of a word**. Those occurring at the first or second letter are more likely to reduce an organism's chances of survival. What's more, a mutation in a start or stop sign is bad news. With no start sign, no protein can be made. With no stop sign, the amino-acid sequence grows indefinitely. As any road user knows, removing a stop sign can have unpredictable consequences!

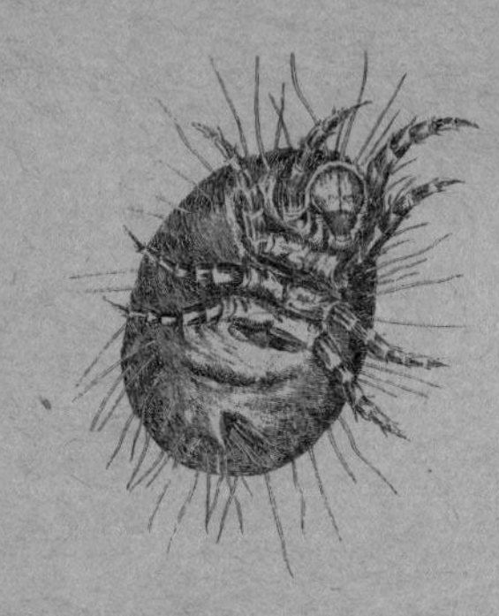

this is probably how most people imagine a bacterium but they are not all bad

Unpleasant Bacteria

Which well-known diseases are caused by bacteria? Try these for size: cholera, tuberculosis, anthrax, syphilis, Black Death, and gonorrhea, to name but a few. Over the last few centuries, all of these party poopers have fallen on hard times, not because of medical advances, but because of increased standards of hygiene.

KEY WORDS

PROKARYOTE: a single-celled organism that lacks a nucleus (i.e. bacteria)

EUKARYOTE: any organism whose cells contain a nucleus (i.e. aardvarks and humans)

BACTERIA

* Now the code had been cracked, the next logical step was to work out the actual sequence of letters in a particular organism's DNA—and to see what proteins were made and how they all fitted together to make a fully grown organism. Simple organisms such as bacteria seemed the best starting point.

SIMPLE YET SURPRISING

* Bacteria are found just about everywhere. You can find them in the deepest oceans, in the soil, on your food, on your face, in your intestines—and even on the pages of this book. Bacteria have a bad reputation because they cause some unpleasant human diseases. But most of them live relatively benign and boring lives, and only a small percentage of them do us any harm.

* Bacteria are not much to look at. They consist of just a single microscopic cell, and one bacterium looks a lot like another. But as biologists were soon to find out, looks can be deceiving—and inside the cell there was a great deal of intrigue. For a start, there was no nucleus. What's more, instead of paired chromosomes, they had just a single circular chromosome made up of double-stranded DNA.

some of my best friends are bacteria

* Bacteria had other surprises in store. As well as the single circular chromosome, which all individual bacteria possessed, some cells carried additional and much smaller circular molecules of double-stranded DNA, called PLASMIDS, which could replicate independently of the rest of the bacterial chromosome.

bacteria are everywhere

* All plasmids carried genes that managed their own life cycles, but some could also affect the characteristics of their host cell. For example, ***many bacteria are resistant to antibiotics—and the genes responsible are usually found on their resident plasmids***, rather than on the chromosome. Plasmids were like genetic lodgers, exploiting the host's resources for their replication and offering genes of their own in return.

Not All Bad

One bacterium, above all others, has been of enormous assistance to scientists engaged in genetic research. It was discovered in 1885, living quietly and happily in human intestines, by German biologist Theodor Escherich and subsequently named in his honor. ***Escherichia coli***—*or* ***E. coli***, *for short—has been the organism of choice for bacterial geneticists over the last 50 years.*

plasmids are also gene carriers

SWITCHING GENES ON

* How was the expression of genes rated? How were genes switched on and off? In the early 1960s three French scientists, Arthur Pardee, Francis Jacob, and Jacques Monod, working at the Pasteur Institute in Paris, made a major breakthrough with the discovery of the operon.

KEY WORDS

OPERON:
a group of adjacent structural genes controlled by a common operator

PROMOTER:
a region of DNA just in front of a gene to which RNA polymerase binds before beginning transcription

"What applies to E. coli applies to E. lephant."
JACQUES MONOD

SUBTLE SWITCHES

* Working with the bacterium *E. coli*, Pardee, Jacob, and Monod had made an intriguing observation. They were growing bacteria on a nutrient medium containing a sugar called lactose, which the bacteria used as an energy source. When the lactose was present in the medium, the bacteria produced two enzymes.

* One enzyme, called PERMEASE, sped up the rate at which the bacteria could take up the sugar from the medium into their cells; and the other enzyme, BETA-GALACTOSIDASE, digested the sugar. However, when lactose was removed from the nutrient medium, the bacteria stopped producing both enzymes.

* It looked as if the two enzyme-coding genes were switched on by the presence of the sugar. But Pardee, Jacob, and

Monod soon discovered that the sugar didn't affect transcription directly. Instead, it worked through a REGULATOR. This was a gene that was permanently switched on and produced a REPRESSOR PROTEIN, which worked like a DNA padlock. When the sugar was absent, the repressor protein prevented the RNA polymerase from making an RNA transcript of the enzyme-coding genes. But the sugar undid the padlock, enabling the enzyme genes to be transcribed.

* The two enzyme-coding genes sat next to each other in the bacterial DNA. In front of these two genes was a PROMOTER—a sequence of a few hundred bases that RNA polymerase uses to anchor itself to

JACQUES MONOD

Monod's quote (opposite) turned out to be not entirely correct. Though there are similarities, gene expression in higher organisms is much more complicated than in bacteria. Unlike bacterial cells, animal and plant cells become specialized to perform different tasks and gene expression relies on a complex preprogrammed pattern of development as well as on external signals.

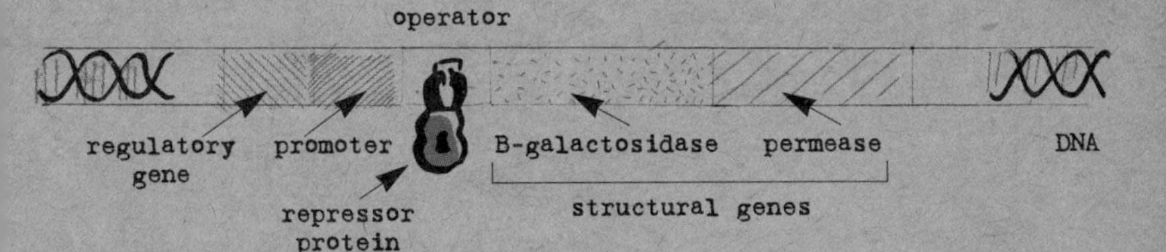

in the absence of lactose, the repressor protein is bound to the control site. Lactose unlocks the padlock and allows RNA polymerase to bind to the promoter and transcribe the two structural genes

the DNA before beginning transcription. But in between the PROMOTER and the two enzyme-coding genes was another site—the operator. It was here that the repressor attached itself to the DNA when the lactose sugar was absent, preventing the RNA polymerase from transcribing the two enzyme-coding genes.

KEY WORDS

STRUCTURAL GENES: code for proteins involved in cell structure and metabolism

REGULATORY GENES: code for proteins which switch other genes on or off

FREE LOVE AMONG BACTERIA

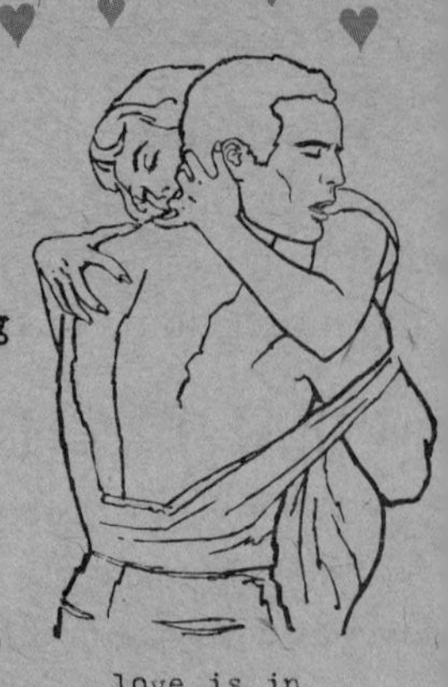
love is in the air

★ Mendel had his peas, Morgan had his fruit flies—but for any aspiring geneticist of the 1960s, bacteria were the stars of the show. And it was the eccentric sex lives of these tiny organisms that were tempting young scientists to turn on, tune in, and pull on a lab coat.

Sex and reproduction

In biology, there is a big difference between sex and reproduction—and it has nothing to do with contraception. Sex is the mixing of genes, while reproduction is the multiplication of individuals. For most organisms, reproduction involves sex. But for bacteria, sex and reproduction are not the same thing at all. Bacterial reproduction is pretty simple stuff. The cell makes a copy of its chromosome, elongates, and then divides in two. This so-called binary fission produces two genetically identical "offspring." When the conditions suit them, bacteria can divide every 30 minutes or so.

SEX AND THE SINGLE CELL

★ Bacterial reproduction is efficient but ultimately kind of dull. By comparison, their sex lives are positively bizarre. By the 1960s bacteria had been shown to lead more varied sex lives than humans. Oswald Avery's transformation experiments 20 years earlier had already shown that necrophilia was an integral part of a bacterium's sexual menu. Avery demonstrated that a bacterium could capitalize on the death of one of its neighbors by ingesting the dead cell's DNA and incorporating it into its own genetic recipe.

★ Transformation was not the only way in which bacteria could exchange genes. If transformation was sex from a distance, then conjugation was sex with a loving embrace. Conjugation began with two bacteria touching and fusing to form a fine channel between them. One cell would then inject some of its DNA—usually a plasmid—through the channel, into its partner. And conjugation could escalate into a full blown orgy, as other cells tried to muscle in on the action.

VIRAL EXCHANGE

* Soon it was discovered that bacteria could also make use of viruses to exchange genes with one another. Occasionally, when new virus particles were being formed inside the bacterial cell, some of the bacterium's own genes were mistakenly incorporated into those of the virus. When the virus went off to infect another cell, it would transmit a mixture of viral and bacterial DNA.

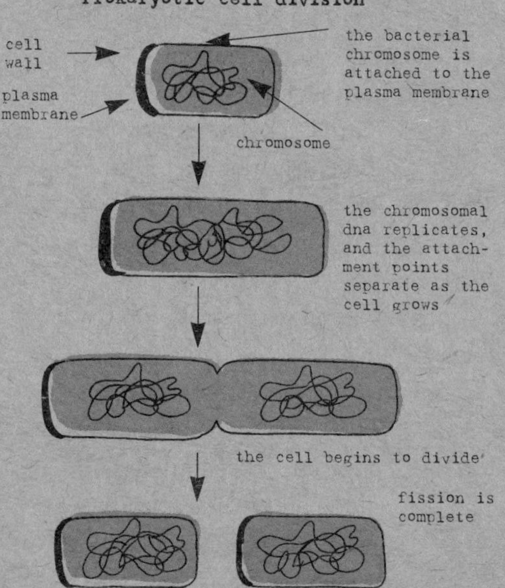

Electrotherapy

Bacterial transformation is a rare occurrence under natural conditions. But scientists soon realized that zapping the bacteria with a high-voltage electric shock made them more than willing to cooperate and take up foreign DNA.

all that's needed is some extra energy

DREAM TICKET

Bacteria were a geneticist's dream. Because they reproduced so fast and were so flexible in their ability to exchange bits of DNA with one another, they seemed like the perfect tool for the study and manipulation of genes.

PULLING OUT THE DNA

* Before geneticists could start fooling around with genes, they had to extract DNA from cells, and then isolate and purify it from all the other cell constituents. This had already become a relatively routine, if ingenious, procedure, relying on an intimate knowledge of the chemical and physical properties of cells and their DNA.

scientists like to snip off mice tails

HACK OFF

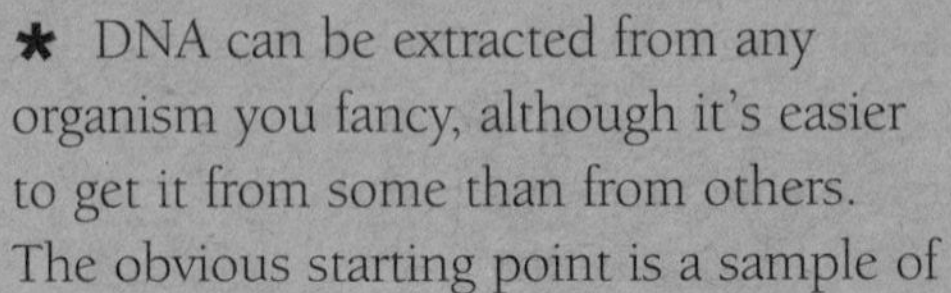

* DNA can be extracted from any organism you fancy, although it's easier to get it from some than from others. The obvious starting point is a sample of cells from the organism that you are interested in. Where these cells come from depends primarily on the size of the animal. In large animals, such as humans and hippopotamuses, a sample of blood contains sufficient cells to start with. In smaller animals, it's sometimes necessary to cut off a bit of tissue. In mice, snipping a piece off the end of the tail is a common way of getting hold of some cells. With even smaller creatures, such as insects, the whole animal is sacrificed in the name of science.

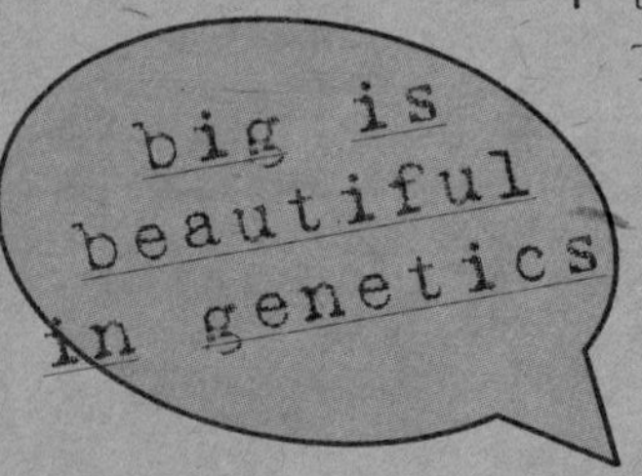

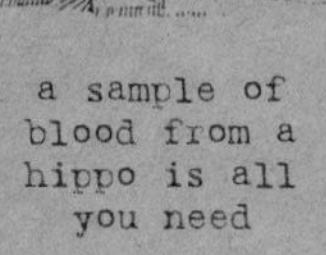

a sample of blood from a hippo is all you need

GOO

RELAX! It's more than just mild

* In multicellular animals, the cells are usually stuck together to form tissues so the sample has to be physically ground up to fragment it. Bacteria present no such problems because they are already in single-celled form.

* Having separated the cells, the next step is to break them open. Cell membranes contain lots of fat molecules, so geneticists use their own version of dishwashing detergent to dissolve the fat in water. A bit of detergent is added to the test tube containing the cells; after a quick shake, you are left with a viscous goo. Floating around in there is the DNA you are interested in.

HEAVYWEIGHT

* Being a long and relatively heavy molecule, the DNA can be separated from the lighter constituents in the goo using the geneticist's favorite toy—the CENTRIFUGE. Once the lightweight stuff has been siphoned off, two chemicals, phenol and alcohol, are used to dissolve away the proteins and RNA, respectively. Eventually, you are left with a tiny and barely visible pellet of DNA stuck to the bottom of the tube. The DNA pellet is dissolved in water and, abracadabra, you have your DNA in user-friendly form.

Centrifuge

A centrifuge is a device used for separating liquids or particles of different densities. It consists of a drum containing holes for test tubes, which rotates very fast (up to 40,000 rpm) in a horizontal circle. As the test tubes rotate, the heaviest particles migrate to the bottom of the tubes, leaving the lighter particles higher up.

a good spin will separate off the DNA

achoo!

viral attacks keep our doctors in business all winter long

MOLECULAR SCISSORS

★ DNA is a very long and unwieldy molecule. So, before geneticists could start manipulating it, they had to find a way of cutting it up into smaller and more manageable chunks. Their search led them, almost inevitably, to bacteria. Bacteria produce their own molecular scissors, which they use to defend themselves from viral attack.

SELF-DEFENSE

★ In 1970, a research team led by **Hamilton Smith** at Johns Hopkins University were studying the bacterium *Hemophilus influenza*. They had noticed that certain chemical extracts taken from the bacterium had the ability to cut DNA molecules at precise points in their sequence. These were the bacterium's RESTRICTION ENZYMES—its very own self-defense system.

★ Bacteria are under constant attack from viruses. Yet very few of these viral assaults result in a successful infection. Restriction enzymes chop up any invading foreign DNA. A chemical shield protects the bacterium's own DNA from being snipped by the molecular scissors.

What's in a name?

Restriction enzymes are named after the bacterium in which they were first discovered. Take ***Eco*RI**, for example. The italicized part of the name is an abbreviation for the bacterium ***Escherichia coli***.

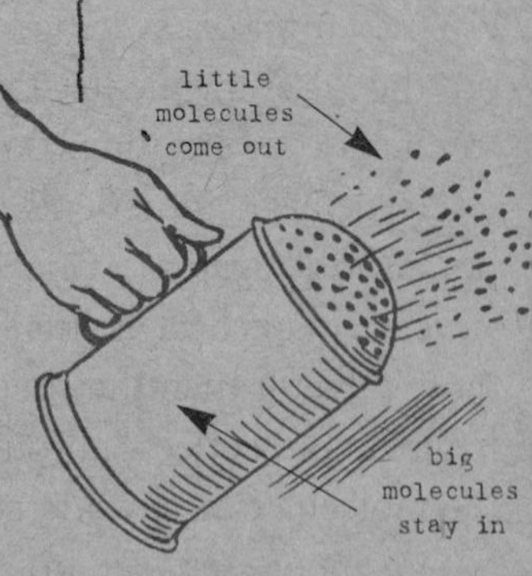

agarose works like a sieve

bring me my bow
of burning gold
bring me my
DNA shield

★ A restriction enzyme will only cut the DNA if it recognizes a specific sequence of letters within it. For example, the restriction enzyme *Eco*RI will only cut the DNA if it detects the letters GAATTC in the DNA sequence. A search of other bacteria species soon uncovered hundreds of different restriction enzymes, each with its own specific recognition site.

★ Once a piece of DNA has been cut by a restriction enzyme, the different-sized fragments can be separated from one another using a technique called ELECTROPHORESIS. DNA molecules have a negative electrical charge. So if an electrical current is applied, they will move toward a positive electrode. The actual separation also depends on the use of a special kind of gel called agarose, which works like a molecular sieve.

★ The mixture of different-sized DNA fragments is put into one end of the gel; when the electric current is switched on, the DNA fragments start to migrate toward the positive electrode at the other end. But because the agarose works like a sieve, the larger fragments find it more difficult than the smaller fragments to squeeze through the molecular meshwork, so the fragments separate from one another in order of size.

LIGHT SHOW

To visualize the fragments, a special stain called ethidium bromide is used, which binds to the DNA molecule. Though colorless in normal light, the stain shines bright orange when viewed under ultraviolet (UV) light.

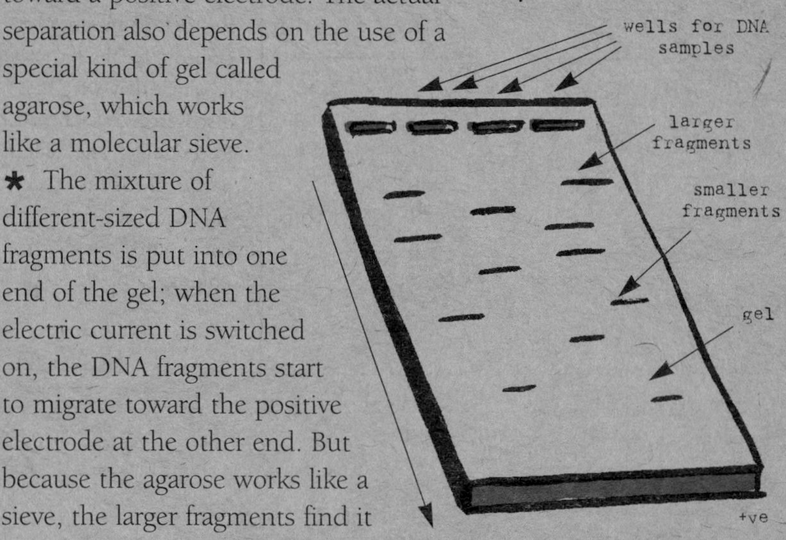

separation of DNA fragments by electrophoresis; each distinct band on the gel consists of thousands of identical-sized fragments

MAKING MAPS

★ Restriction enzymes enabled biologists to start making the first physical maps of DNA. These maps pinpointed the order of restriction sites on the chromosomes, rather than the order of genes. But restriction sites were useful genetic landmarks. By looking for associated patterns of inheritance between these genetic markers and particular characteristics, biologists could begin to trace the location of specific genes.

THE FRAGMENT FACTOR

★ To illustrate how restriction enzymes can be used to make physical maps of DNA, imagine you are interested in making a map of a piece of DNA that is 10,000 letters long. First, you try cutting the DNA with a single restriction enzyme—*Eco*RI, for example—that recognizes and cuts the DNA wherever the letters GAATTC occur in the DNA sequence. After incubating your DNA with the enzyme for a couple of hours, you run the DNA out on an agarose gel.

★ When you come to look at the agarose gel under UV light you see that the original DNA has been cut into three smaller fragments, whose lengths are about 2,000, 3,000, and 5,000 letters respectively. This immediately tells you that your original 10,000-letter DNA has two cut sites—two places where the letters GAATTC occur in the sequence.

KEY WORDS

PHYSICAL MAP: map based on the order of DNA letters
GENETIC MAP: map based on the order of genes (see page 55)

Easy?

On the right is a relatively simple example! In practice, biologists may use many more enzymes, either separately or combined, to work out the ordered arrangement of restriction sites.

★ What you don't know yet is the order in which these cut sites occur. With two cut sites, there are three possible arrangements:

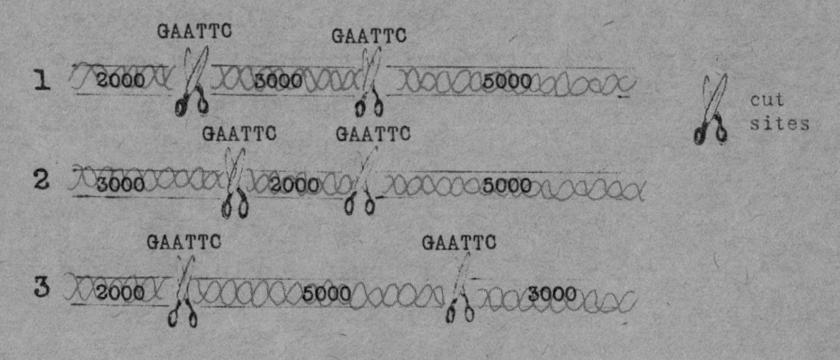

★ To decide between these alternatives, a second enzyme is needed. Suppose a second enzyme (*Bam*HI, for example) which has the recognition site GGATCC cuts the 10,000-letter DNA into just two fragments, one of 500 letters and the other 9,500 letters. With only one cut site there is only one possible arrangement:

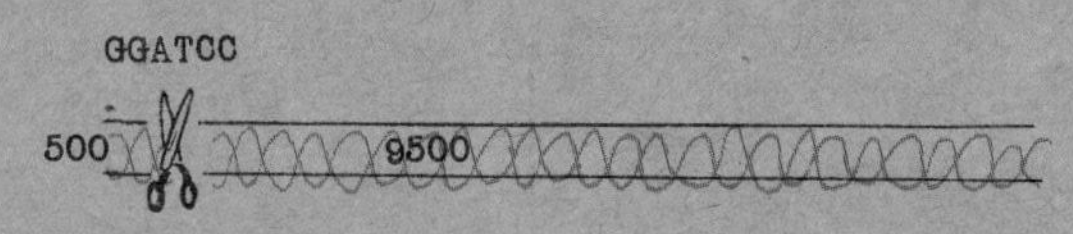

★ But using a mixture of the two enzymes simultaneously can help resolve the order of the *Eco*RI restriction sites. If a double digest produces four fragments consisting of 500, 2,000, 2,500 and 5,000 letters, then we know immediately that the second arrangement describes the correct order of restriction sites for *Eco*RI:

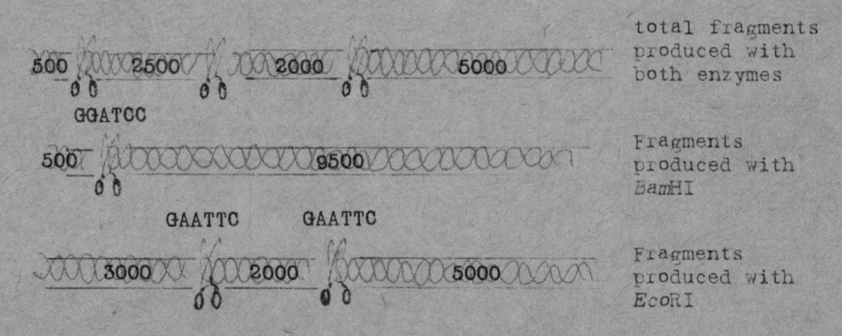

COOKING LESSONS

★ Genetics was beginning to resemble advanced cooking, with DNA as the main ingredient. The laboratory had become the geneticist's kitchen, complete with expensive chemicals and gadgets. But the potential of this technology was enormous. You could potentially extract DNA from any organism you wanted, chop it up, and use bacteria as growth factories to produce millions of copies of individual genes for further analysis.

Today we are going to make E-coli. Bacteria are extremely easy to make in the kitchen. A dirty plastic plate is all it takes to make a lovely collection of bacteria for your family ... next week penicillin

DNA soup?

CLONING GENES

★ The amount of DNA obtained from a simple DNA extraction was hardly sufficient for most experimental purposes. If geneticists wanted to mess around with genes, they needed to produce them in millions. So before a gene could be studied in any detail, it needed to be CLONED.

★ Cloning involved the generous, if not entirely willing, participation of bacteria. After the DNA from the chosen organism had been extracted and purified, restriction enzymes were used to cut up the DNA into lots of smaller and more manageable fragments. To get each of these fragments into a bacterium required a VECTOR. Plasmids are ideal for this purpose. With some friendly chemical persuasion, plasmids could usually be encouraged to incorporate an individual DNA fragment into their own genetic recipe. The plasmid, now containing a bit of foreign DNA, could then be moved back into its bacterial host.

harvesting the DNA

GROWING AND HARVESTING

★ Bacteria like *E. coli* are extremely easy to grow in the laboratory. A plastic plate measuring about 3 inches (10cm) in diameter and containing a mixture of a few basic bacterial nutrients is more than enough to keep them happy. The multiplying potential of bacteria and of the plasmids themselves ensures that after a few days your plastic plate contains millions of copies of your original DNA fragments, tucked away inside the bacterial hosts. All that then needs to be done is to "harvest" the DNA from each of the plasmids within the bacterial cells.

KEY WORDS

CLONES AND CLONING:
A clone is an identical copy. To clone means to make one or many copies of something. In biology, the term clone can refer to an organism as well as a gene. Identical twins, for example, are genetic clones of one another.

GENE LIBRARY:
a collection of cloned DNA fragments that represents the entire genetic recipe of an organism; also called a gene bank

RECOMBINANT DNA:
DNA that contains genes combined from two or more distinct sources by means of genetic engineering

VECTOR:
any vehicle used to insert a foreign piece of DNA into a host cell; in genetic engineering, vectors are usually plasmids or viruses

sorting out the sequence

SEQUENCING DNA

* Having obtained millions of copies of a particular fragment of DNA, the next obvious task was to sequence it to determine its linear order of letters.
DNA sequencing was first perfected in 1977 by the Cambridge biochemist Fred Sanger.

FRED SANGER

Fred Sanger was the first person to deduce the amino-acid sequence of a protein, for which he received a Nobel prize in 1958. In 1980, he received a second Nobel prize for his pioneering work on DNA sequencing. It was the first time that two prizes in chemistry had been awarded to the same person. For Sanger, sequencing was obviously in his genes! In Britain, the Sanger Centre in Cambridge is the home of DNA sequencing, not only in humans but in many other species.

RIGHT, SAID FRED

* Sanger's DNA sequencing technique took its lead from the way in which a new DNA strand was made from a template strand during DNA replication. With a few important modifications, he simply transported the chemical reactions of DNA synthesis from the nucleus into the test tube.

* Sanger took four test tubes, which he labeled A, G, C, and T. Into each tube he put multiple copies of the stretch of DNA to be sequenced, plus a surplus of each of the four DNA letters (A, G, C, and T). Next he added some chemically modified versions of the DNA letters, which were to act as inhibitors during the DNA sequencing reaction. Into the tube marked A he added the A inhibitor; into tube G he added the G inhibitor; and so on for the other two tubes. Next, he heated up the mixtures in each tube, which caused the double-stranded DNA fragments to completely unzip, leaving lots of single-stranded DNA.

* After lowering the temperature, Sanger added some DNA polymerase to each tube, which helped to pair up the freefloating DNA letters with their complementary bases on the DNA templates. If a complementary inhibitor was incorporated, rather than a complementary letter, then any further addition of letters to the growing DNA chain was blocked. At the end of the sequencing reaction, tube A, for example, would contain a mixture of subfragments whose lengths corresponded to the position of the A's on the original template.

Sanger developed a new DNA sequencing technique

CONFUSED?

Imagine if the sequence of the DNA fragment you are trying to work out is ATGCCTAGGC. The fragments produced in each of the four tubes would be:

Tube **A**	Tube **G**	Tube **C**	Tube **T**
A	AT**G**	ATG**C**	A**T**
ATGCCT**A**	ATGCCTA**G**	ATGC**C**	ATGCC**T**
	ATGCCTAG**G**	ATGCCTAGG**C**	

Because Sanger's four reaction tubes generated all the possible sub-fragments of his template sequence, he could work out the original sequence simply by separating all the different-sized fragments, using electrophoresis, and then "reading" the sequence off the gel.

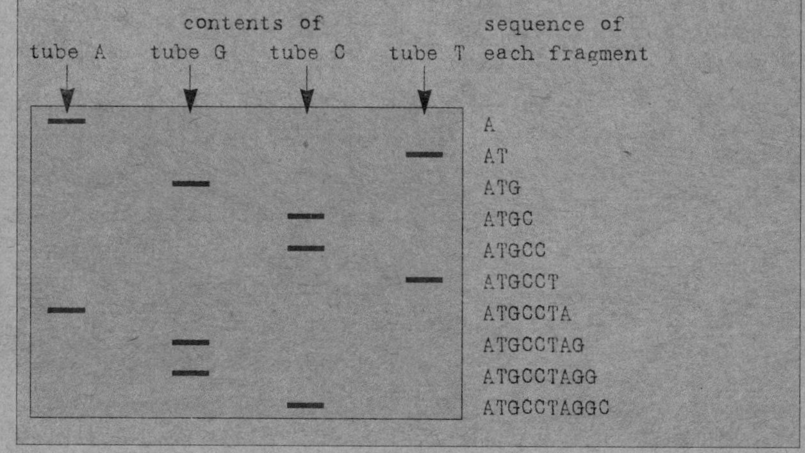

surprise! surprise!

DNA SURPRISES

★ Sanger's discovery prompted much rejoicing among the genetic community. Now it would at last be possible to read the DNA instruction manual of an organism from beginning to end, and maybe even to come to grips with the big question of how a virus, a bacterium, a plant, or even a human, is put together. DNA sequencing was a radical new technique, and geneticists took to it in droves. But there were more surprises in store.

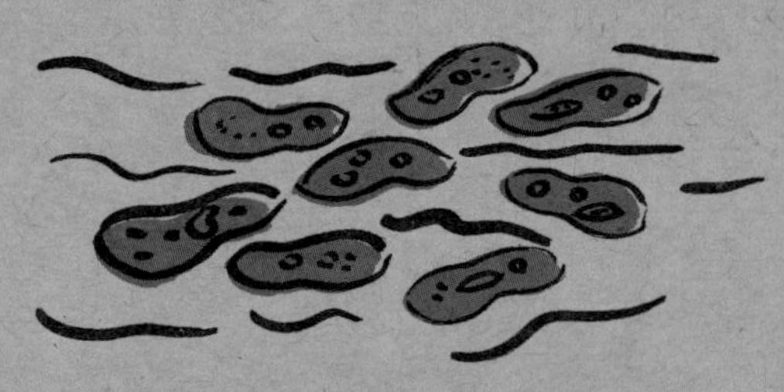

VIRUSES AND BACTERIA

★ The first organisms to come under the sequencer's scrutiny were viruses and bacteria. They seemed simple enough. Their genes were arranged nicely and neatly in a regular linear order, and each gene made a piece of RNA that was translated into a protein. But when geneticists turned their attention to more complex organisms, things began to look a little more complicated.

Introns and Exons

Remove the "introns" from the following English sentence to reveal the meaning of the gene:
my*usxvcf*petaar*zftyey*dvark*sqd*is *ccr*called*gyios*nigel

KEY WORDS

INTRONS:
the noncoding regions of a gene

EXONS:
the coding regions of a gene; introns and exons are only found in eukaryotic cells (cells containing a nucleus) and not in bacteria

PSEUDOGENE:
a DNA sequence which resembles that of a normal working gene, but contains mutations that prevent it from functioning

EXCESS BAGGAGE

★ Some genes had far more DNA than was needed to make a protein. It was as if they were carrying surplus DNA baggage. Genes, it was discovered, were made up of CODING SEQUENCES (EXONS) interspersed with NONCODING SEQUENCES (INTRONS).

scientists celebrated Sanger's discovery

★ Although the entire DNA sequence of the gene was transcribed into an mRNA molecule, the introns were edited out before the mRNA transcript left the nucleus for the protein factories. And noncoding sequences were not turning up only within genes…

some genes carry surplus baggage

Pseudogenes

Often, the genes themselves were separated by huge stretches of meaningless DNA. This DNA seemed to make no protein, no mRNA, nothing. It was just sitting there, taking up space on the chromosomes. Some stretches at least resembled the coding genes they sat next to. Perhaps these so-called pseudogenes did once form part of the genetic recipe—but riddled with mutations, they had become defunct and were now no more than genetic fossils?

jumping genes arrive uninvited

JUMPING GENES

* Most of the genetic "junk" seemed to consist of highly repetitive DNA sequences. Humans, for example, had between 300,000 and 500,000 copies of a 300-letter stretch of DNA dispersed within their genetic recipes. Geneticists would soon discover that an organism's genome (its complement of DNA) included not only its own genes but a whole host of genetic parasites, as well.

SELFISH DNA

* There had long been a sneaking suspicion among scientists that not all DNA acted in the organism's own self-interest. As early as the 1950s, the geneticist **Barbara McClintock** had postulated the existence of so-called "JUMPING GENES," which could move from one region of a chromosome to another. RECOMBINANT DNA technology and DNA sequencing had pulled the lid off many of the genome's secrets, and jumping genes—or TRANSPOSABLE ELEMENTS, to give them their proper name—began to turn up all over the place. Usually about 1,000 or so letters long,

jumping genes were first investigated in corn

BARBARA McCLINTOCK

The American geneticist **Barbara McClintock** (1902–92) first presented her pioneering work on jumping genes in corn in 1951. But the scientific community was not convinced—and it wasn't until the 1970s, when molecular biologists confirmed the existence of jumping genes in bacteria, that her work got the credit it deserved. She was awarded a Nobel prize in 1983.

Did you know that

that about 75% of your DNA is junk?

jumping genes relied on the host's replication system to make copies of themselves. ***Jumping genes, it was discovered, coded for transposase, an enzyme that enabled them to hop around on the DNA.*** Unlike the bacterial plasmids, there was nothing the slightest bit altruistic about these mobile genetic elements. They were molecular parasites whose sole *raison d'être* was to replicate and transmit themselves to the next generation.

NOMADIC NUISANCE

* For an organism, jumping genes are definitely a genetic liability. They can cause major mutations if they insert themselves into one of an organism's own genes. In fact, it is now thought that many mutations are caused by these chromosomal nomads.

* Jumping genes could also explain the existence of genetic junk. The vast majority of JUNK DNA is now thought by scientists to represent the remains of jumping genes and other genetic parasites, such as viruses, which inserted themselves into chromosomes years ago, but have since been rendered immobile by mutation.

so much genetic garbage

Genetic Garbage

It is estimated that about 75% of the DNA in humans is junk. In other species, the figure is even higher. Amphibians seem to be the biggest hoarders of genetic garbage. Salamanders, for example, have 20 times more DNA than humans, and most of it is junk.

EXTRA-NUCLEAR AFFAIRS

★ DNA is not confined to the nucleus. In both plants and animals, genes are also carried on independent circular bits of DNA in specialized cellular structures outside the nucleus.

REVISED RELATIONSHIP?

The close similarity between the circular DNA molecules of chloroplasts and mitochondria, and the circular chromosomes of bacteria, has led many scientists to speculate on whether mitochondria and chloroplasts were once parasitic bacteria that have since developed a more mutually beneficial relationship with their tenants.

POWERHOUSES

★ The region of a cell between its outer membrane and its nucleus is anything but a cell wasteland. It's jammed with ORGANELLES—the cell's miniature equivalent of the body's organs. Each type of organelle has its own job to perform to keep the cell ticking. The most common organelles are the sausage-shaped MITOCHONDRIA. *These are the powerhouses of the cell, the places where we ultimately get all our energy from.*

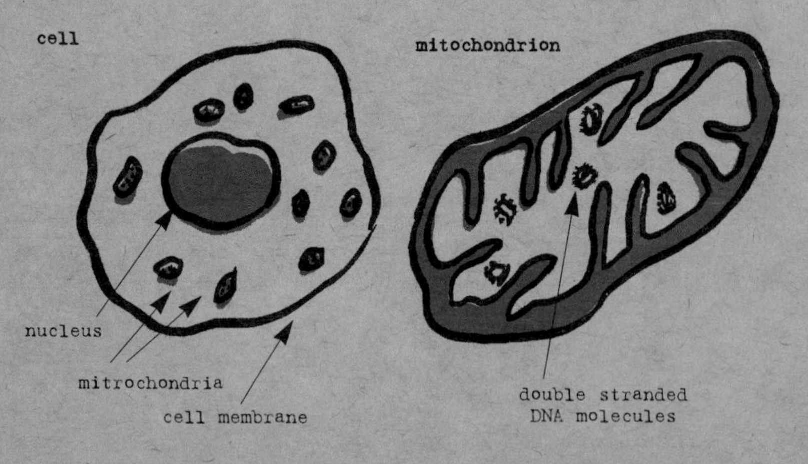

The oxygen that we breathe into our lungs eventually makes its way, via the blood stream, inside the mitochondria in our cells, where it is used to "burn" fuel (food) to produce energy.

★ Mitochondria contain circular pieces of double-stranded DNA that carry only a few genes—13 in humans. Oddly, these genes are not all associated with the mitochondrion's function. Most of the proteins that the mitochondria need to burn fuel are coded for by genes in the cell's nucleus.

★ Another odd thing about MITOCHONDRIAL DNA (mtDNA) is its strange pattern of inheritance: ***it is passed on exclusively through the maternal line***. Whether you are male or female, your mtDNA always comes from your mother. What's more, it never shuffles its genes with any of its neighbors. Mutations are the only way in which mtDNA can change from one generation to the next.

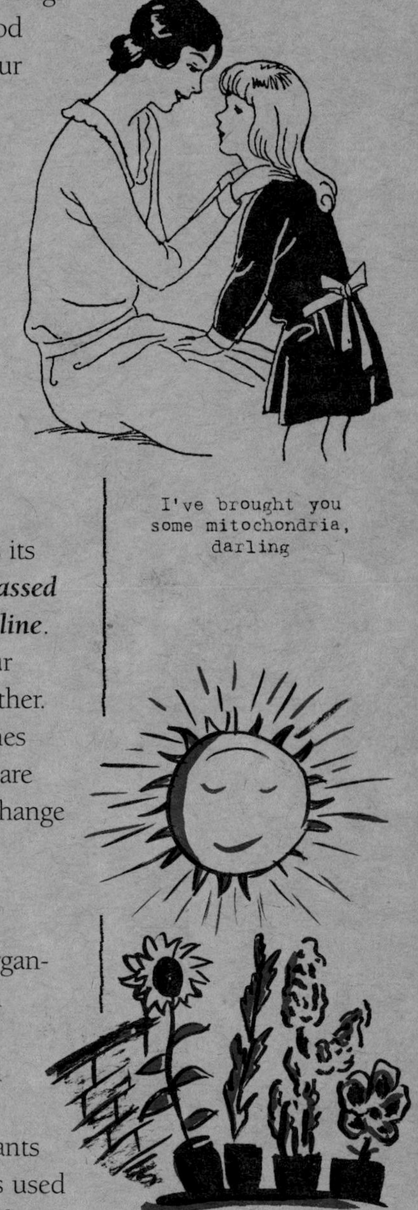

I've brought you some mitochondria, darling

plants like sunshine

GOING GREEN

★ Mitochondria are not the only organelles that contain their own DNA. In plants, CHLOROPLASTS also have a circular molecule of double-stranded DNA. Chloroplasts contain the pigment chlorophyll, which gives plants their characteristic green color and is used by the plant to convert energy from the sun into food (sugar).

CHANGE OF DIRECTION

★ Rules in science are made to be broken. In the 1960s, James Watson's "central dogma" of molecular biology went the way of most sweeping statements: it was swept away after a remarkable discovery by Howard Temin **and** David Baltimore.

HIV AND AIDS

AIDS (Acquired Immune Deficiency Syndrome) is caused by the human immunodeficiency virus (HIV). This retrovirus inserts itself into the body's T-cells, which form part of the immune defense system. AIDS-related deaths are normally the result of "opportunistic" infections caused by bacteria, fungi, and other viruses, which are able to exploit the weakened immune system.

AGAINST THE FLOW

★ Watson's central dogma, which is still espoused by biology teachers up and down the land, stated that genetic information always flowed in one direction—DNA makes RNA makes protein. To be fair, this is generally true. But in 1964 Howard Temin found an important exception to the rule. Temin was working with a weird virus that had RNA, rather than DNA, as its genetic material.

research into the AIDS virus is ongoing

THE RETRO LOOK

★ Temin found that the virus was unable to infect a host cell if he added a chemical inhibitor of DNA synthesis or transcription. This immediately suggested that RNA viruses needed to synthesize DNA in order to copy themselves. In other words, ***Temin discovered that there was a reverse flow of genetic information.***

Retroviruses are more than just eccentric genetic oddities. They are responsible for a number of serious human diseases, including AIDS and certain cancers. Because they have extremely high mutation rates and can evolve rapidly, effective treatment is difficult–it's like trying to hit a moving target.

* In 1970 Temin and Baltimore discovered that the RNA viruses had a special enzyme which could use RNA as a template for the synthesis of a complementary DNA strand. They called the enzyme REVERSE TRANSCRIPTASE, and the viruses that contained it were later christened RETROVIRUSES—after their ability to reverse the usual direction of information flow.

RETROVIRUS INVASION

* After a retrovirus has entered a host cell, the reverse transcriptase enzyme synthesizes a DNA strand complementary to the RNA template. The single-stranded DNA then makes a complementary strand, and the double-stranded DNA inserts itself into one of the chromosomes of the host. The DNA is then transcribed into single-stranded mRNA molecules.

* Some of these move off to the cell's protein factories to be translated into viral proteins. The remainder are encased in these proteins to form new virus particles. These viruses are now ready to burst out of the host cell and repeat the whole unpleasant process in another unsuspecting cell.

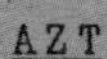

AZT

Though there is, as yet, no cure for AIDS, use of azidothymidine (AZT) does seem to increase the survival time of AIDS patients. AZT works by blocking DNA synthesis, so that the HIV retrovirus is unable to make copies of itself.

FINDING GENES

★ How do geneticists actually go about finding specific genes? How do they start? In human genetics, most of the research in this area has concentrated on genes that cause disease, the approach combining aspects of traditional and modern genetics. Family pedigrees can provide important information on a gene's pattern of inheritance, and modern genetic technology is used to refine the search down to specific regions of a chromosome.

NEEDLES AND HAYSTACKS

★ Despite the huge advances made in genetic technology over the last 30 years or so, searching for a gene is still a bit like looking for a needle in a haystack. Nevertheless, many disease-causing genes have been identified and pinned down to specific locations on chromosomes. One of the most celebrated instances of gene hunting was the search for the cystic fibrosis gene in the late 1980s.

★ The first job was to look at the inheritance pattern of the disease, using family case histories. Cystic fibrosis followed a Mendelian pattern of inheritance, which told the researchers immediately that they were looking for a single gene. The disease was no more common in males than females, which excluded the X chromosome from the search.

LUNG SUFFERERS

Cystic fibrosis is a single gene defect whose major symptom is an abnormal build-up of mucus in the lungs. The disease affects about 1 in 2000 people.

looking for needles in haystacks

GENE JUNCTION

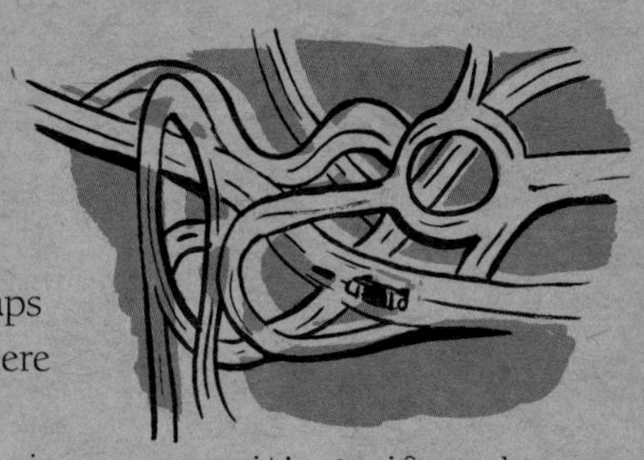

it's as if you have lost your car...

★ By the late 1980s the human chromosomes had been extensively mapped. These were not so much maps of genes as maps of restriction sites—the places where restriction enzymes cut the DNA.

★ Imagine the 23 chromosome pairs to be like 23 different highways. You have lost your car, but you know that the car, the metaphorical cystic fibrosis gene, is somewhere on one of those 23 highways. The restriction sites are like mile markers on the highways. Each marker gives you two pieces of information: it tells you which highway you are on, and where you are in relation to the other mile markers on the highway.

★ So the task facing the geneticists was to identify the "mile markers" that were physically closest to the cystic-fibrosis gene (once you find the mile marker nearest to your car, you've effectively found the car). Because segments of DNA that are close to one another on the chromosome are inherited together, the search involved looking at all the markers until they found the ones whose pattern of inheritance was most closely associated with that of the disease.

★ It was a case of trial and error, but eventually, the gene was pinpointed. The stretch of DNA between the two closest site markers was cut out and cloned, and by 1990 the gene had been sequenced.

Long and Winding Road

After years of searching, the cystic fibrosis gene was finally tracked down to a region of chromosome 7. It is an unusually large gene, made up of about 250,000 DNA base pairs. Only about 2 percent of the gene is actual coding DNA. The rest is made up of noncoding intervening sequences (introns). The normal gene codes for a protein 1,480 amino acids long. Most cystic fibrosis cases are caused by a small mutational deletion in the gene, which leads to a defective protein.

CHAPTER 4

USING OUR KNOWLEDGE

★ Over the last few years all sorts of new technological tricks have emerged in molecular genetics. One of the most significant of these has been the Polymerase Chain Reaction (PCR). PCR is the geneticist's equivalent of the photocopier. Starting with just a few molecules of DNA, it can generate literally billions of copies of a specific DNA sequence in a few short hours.

Voyage of Discovery

PCR was invented by American geneticist Kary Mullis, who was awarded a Nobel prize for his work. He claimed the idea came to him after a drug-induced hallucinatory trip into a DNA molecule!

PCR AND PRIMERS

★ ***PCR is a quick and simple way of cloning genes in a test tube***, without the need for bacteria, and has replaced conventional cloning techniques in many areas of molecular genetics. It does have its limitations, however: it can only be used when you already have some sequence information on the bit of DNA you wish to clone. Otherwise, it's back to bacteria.

★ Every PCR reaction requires a pair of DNA PRIMERS. These are short stretches of single-stranded DNA that are complementary to the DNA sequences flanking the gene which is to be copied. These primers go into the ubiquitous test tube along with a sample of DNA, generous quantities of each of the four nucleotides (A, G, C, and T), and some DNA polymerase enzyme.

THREE EASY STEPS

* The PCR reaction consists of a cycle of three steps repeated again and again. Each step lasts about a minute. In the first step, the reaction mixture is heated up to about 194°F (90°C), which separates the double-stranded DNA into two single strands.

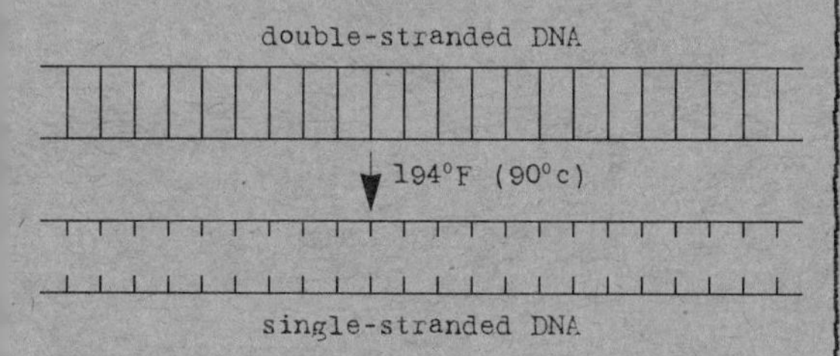

* Although the primers are sitting in the test tube, they cannot bind to the DNA at such a high temperature. So in the next step, the temperature is lowered to around 122°F (50°C), which allows them to attach themselves to the complementary sequences on each of the single strands.

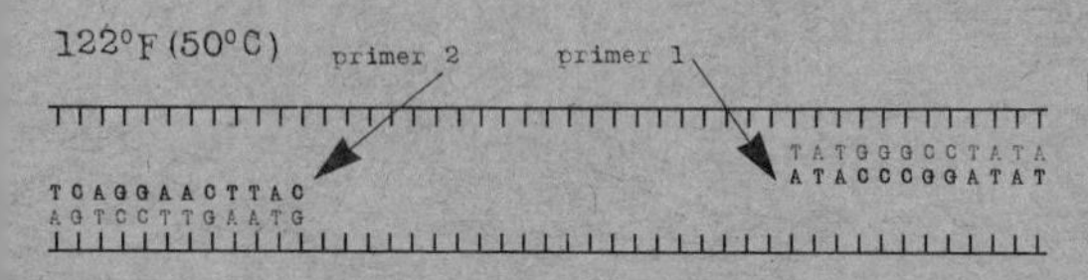

* In the final step, the temperature is increased to about 162°F (72°C). At this temperature, the DNA polymerase begins to synthesize new DNA strands, starting from each of the two primer sequences. Repeated cycles of the three steps produce a doubling of the DNA molecules each time. After about 30 cycles, the number of DNA copies will have increased about a billion times.

Thermocycler

PCR reactions are performed in a thermocycler—a machine that can rapidly shift between temperatures in a programmed order.

Hot Stuff

Ordinary DNA polymerase won't work at the high temperatures involved in the PCR reaction, so a special heat-resistant variety is required. This is derived from the bacterium ***Thermus aquaticus,*** *which lives in hot springs at temperatures of up to 194°F (90°C).*

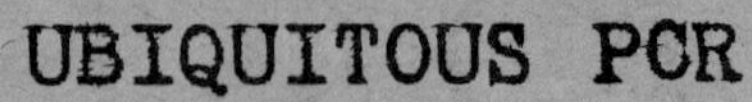

UBIQUITOUS PCR

★ PCR has many applications in modern genetics—but it has become particularly important in the clinical diagnosis of genetic disease and in the field of forensic science.

Effective range of PCR

PCR works best on segments of DNA that are longer than 100 base pairs and shorter than about 3,000 base pairs. The technique can be used on as little as a picogram of starting DNA–i.e. 0.000000000001 grams of DNA!

GENETIC SCREENING

★ It is now common practice to screen the genetic recipes of prospective parents and fetuses for disease-causing genes, particularly when there is a family history of an inherited disorder. This is possible thanks to PCR. ***By using primers specific only to sequences found in the disease-causing gene, a PCR test can reveal whether an individual carries the gene.***

★ If the gene is present, the primers will bind to their complementary sequences on the DNA molecule, and the PCR reaction will produce millions of copies of the defective gene. These are detected by electrophoresis and ethidium-bromide staining of the DNA. If an individual does not carry the gene, the primers will have no complementary sequence to bind onto, giving a blank result.

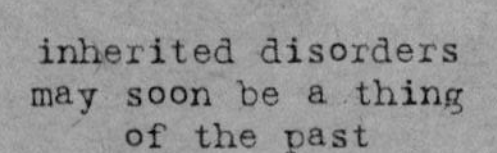
inherited disorders may soon be a thing of the past

a single cell left behind can incriminate

Criminal copies

One of PCR's great strengths is that it works with even the smallest amount of DNA. For this reason, it has become an invaluable tool in forensic science. Crime scenes are full of potential genetic clues. Criminals often unwittingly leave behind small amounts of hair, skin, blood, and saliva. All these bodily tissues are made up of cells that contain copies of the criminal's genetic recipe. Although DNA can be extracted from a single hair follicle, the amount obtained is so minute as to be apparently undetectable—that is, until PCR is used to produce sufficient gene copies for a detailed genetic analysis.

forensic science has made leaps and bounds since PCR was invented

DIAGNOSIS

* **PCR is used in a similar way as a diagnostic test for the presence of disease-causing organisms such as the HIV virus and *Mycobacterium*, the bacterium that causes tuberculosis.** In an HIV test, DNA is extracted from a small sample of the patient's blood. Then a PCR reaction is carried out using primers specific to sequences found only in the genes of the HIV virus. If the primers do not find a sequence to bind onto, the virus cannot be present.

women are told about genetic risks before having a baby

CARRIERS AND COUNSELORS

★ Nowadays PCR is used to test whether prospective parents are carriers for some of the more well-known and serious genetic diseases. The results of such a test can help couples make informed decisions about whether or not to have a child and the genetic risks involved.

Lethal genes

Each one of us carries about five or six recessive genes in our genetic recipes that would kill us if we had two copies instead of one. However, the appearance of such lethal recipes in a child depends on two carriers of the same gene meeting and both passing their recessive gene on to their offspring.

CYSTIC FIBROSIS

★ Cystic fibrosis is caused by a defect in a protein whose job, in normal individuals, is to maintain a chemical equilibrium across the membranes of cells. People who suffer from it experience an abnormal build-up of thick mucus in their respiratory tract and digestive system. Apart from causing breathing and digestion problems, the mucus is a perfect breeding ground for many species of dangerous bacteria, so infections such as pneumonia are common.

KEY WORDS

CARRIER: a person who carries a single copy of a recessive disease-causing gene

GENETIC COUNSELOR: a counselor trained to educate prospective parents about genetic disease and the risks for their children

HYPOTHETICALS

★ Imagine a situation in which a couple are planning to have a child. Both are perfectly normal, but each of them comes from a family that has a history of cystic fibrosis. There is a strong possibility that the parents both carry a single copy of the recessive cystic fibrosis gene. In such instances, they are likely to be referred to a genetic counselor who is trained to educate people about all aspects of genetic testing, and to help them understand any potential risk of genetic disease for their children. In this example, the genetic counselor would almost certainly recommend a genetic test for the cystic fibrosis gene. Of course, the decision as to whether to have the test ultimately rests with the couple themselves.

THE TEST

* A small sample of blood or a simple mouth swab is taken from both people to collect some cells. The DNA is extracted from the cells, and a PCR test using primers specific to the cystic fibrosis gene will reveal whether or not either or both of them are carriers.
* If neither or only one of them is a carrier, then there is no chance that their child could inherit the disease. Because cystic fibrosis is a recessive disease, an individual must inherit two copies of the cystic fibrosis gene, one from each parent, to get the disease. If both parents are carriers, then their child would have a one in four chance of inheriting it.

Tay-Sachs syndrome

This is an extremely unpleasant recessive genetic disease that results in blindness, severe mental retardation, and death before the age of five. In the U.S., the disease is common among Ashkenazi Jews. Arranged marriages are practiced in some Ashkenazi communities, and a genetic test can determine whether partners are genetically suitable for one another.

information is the key

DECISIONS AND DILEMMAS

an impossible choice

★ Sometimes the decision about whether to be tested for a specific genetic disease can be an extremely difficult and traumatic one. This is particularly true of Huntingdon's disease, which does not become apparent until middle age. The results of such a test will not only reveal the possible fate of a child, but also the fate of the parent.

HUNTINGDON'S DISEASE

★ Huntingdon's disease is a dominant genetic disease. If you inherit a single copy of the Huntingdon's gene, then you will get the disease. But until middle age anyone carrying the Huntingdon's gene appears to be perfectly normal. At some time between the ages of 35 and 50 the gene becomes active and the symptoms of the disease begin to appear.

is it better to know or to remain ignorant?

The Tiresias Complex

The dilemma confronting someone faced with the option of a test for Huntingdon's disease has been termed the Tiresias complex. In Greek mythology, the seer Tiresias confronted Oedipus with the dilemma, "It is but sorrow to be wise when wisdom profits not."

Woody Guthrie

The great American folk singer Woody Guthrie (1912–67), who was a major influence on Bob Dylan, was just one among several members of his family to die of Huntingdon's disease.

* These symptoms make unpleasant reading: severe mental and physical deterioration, uncontrollable muscle spasms, personality changes, insanity, and, ultimately, death.

* The gene for Huntingdon's disease was cloned in 1993, and there is now a simple

at present there is NO CURE for Huntingdon's disease

test available to diagnose whether a person carries the gene. But anyone who suspects they have the gene faces an awful dilemma. Do they have the test and reveal their fate? Or do they remain in the dark and trust their luck?

* Many people would go for the latter option. But if a person does have the disease and doesn't take the test, they may unknowingly pass the gene on to their children before the symptoms become apparent.

BUT...

* The only note of optimism in this nightmare of choices is that an early diagnosis of the Huntingdon's gene might contribute to the development of an effective treatment for the disease.

GEORGE HUNTINGTON

In 1872 **George Huntington** wrote about a hereditary defect "which exists so far as I know almost exclusively on the east end of Long Island." The ancestry of the disease was later traced to two brothers in Suffolk, England. It's ironic that the name of a disease caused by a mutation has itself gone through a mutational change. Due to a published misspelling in 1893, which went unnoticed, Huntington became Huntingdon and has been saddled with alternative spellings

the first known case of Huntingdon's was in Suffolk, England

EARLY-WARNING SYSTEM

* Human embryos, as well as prospective parents, are now commonly screened for disease-causing genes and other genetic defects. If a genetic test reveals that child would have a short and painful life, then a mother may consider abortion as the most humane option. But not all genetic diseases are fatal, and early diagnosis of some diseases can increase the chances of successful treatment.

PRENATAL SCREENING

Prenatal diagnosis is normally only offered to older mothers or those who have a family history of inherited disease. If it is likely that termination may be advisable, then clearly early testing is desirable. Unfortunately, neither amniocentesis nor chorionic villus sampling are ideal in this respect: cell samples cannot be taken until about 16 weeks and 10 weeks, respectively.

TAKING SAMPLES

* All kinds of information can be obtained about the genetic health of a developing embryo from just a small sample of its cells. The chromosomes can be looked at under a microscope to determine its sex and to check whether there is the normal chromosome complement of 23 pairs.

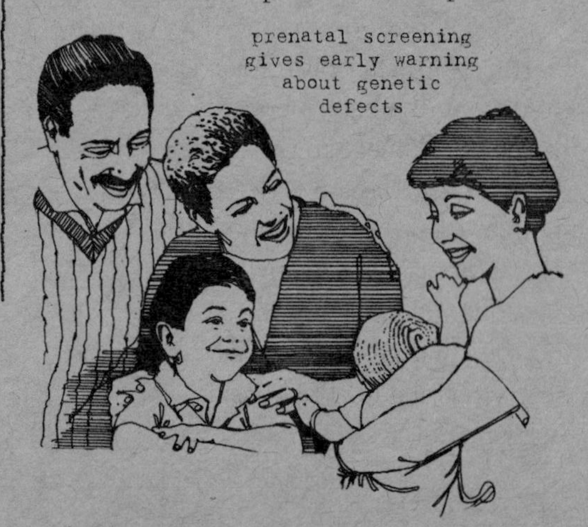

prenatal screening gives early warning about genetic defects

About 50 genetic defects can be diagnosed by looking at the genes themselves (i.e. the DNA) or their protein products.

* There are two ways of obtaining embryonic cells from the mother's womb. In AMNIOCENTESIS, a sample of the amniotic fluid is taken from the mother's womb. This will contain cells shed from the embryo's skin tissue. In the alternative method, known as CHORIONIC VILLUS SAMPLING (CVS), a sample is taken from some of the embryonic cells that are destined to form part of the placenta.

SUCCESS STORY

* One of the most remarkable success stories of early diagnosis concerns the recessive disease PHENYLKETONURIA (PKU). Left untreated, PKU causes severe mental retardation. Babies born with the disease appear perfectly normal, but they are unable to produce an enzyme that metabolizes the amino acid phenylalanine. In the womb, the baby can survive on the mother's own enzyme—but after birth toxic chemicals build up in the baby's body, leading to irreversible damage to the central nervous system. If the disease is detected early enough, then the baby can be put on a special diet lacking phenylalanine; consequently no symptoms develop and the child grows up to be normal. Early screening for PKU has saved thousands of lives.

Tests

At the moment, tests exist for only a small percentage of the 5,000 or so known genetic diseases; and of those that can be detected, only a fraction can be successfully treated. Nevertheless, for the diseases that are treatable, early diagnosis can be important.

early diagnosis offers the best hope

BEFORE THE WOMB

★ There is now a new form of prenatal screening, called preimplantation diagnosis, which can benefit women who have a family history of a serious genetic disease and for whom abortion is either unacceptable or medically inadvisable.

the sperm and egg are fused together in a test tube

TEST TUBE BABIES

About 10% of all couples experience problems in achieving conception. In vitro fertilization (IVF) is just one of many infertility treatments that have been developed over the years. The first "test-tube baby" was born in 1978, in England. Since then, thousands more have been conceived.

thousands of babies have been conceived using IVF

TEST TUBE FUSION

★ Preimplantation diagnosis screens an embryo for genetic defects before the embryo attaches itself to the mother's womb. This is only possible when the embryo has been produced by IN VITRO FERTILIZATION (IVF)—in other words, when sperm and egg have fused in a test tube, rather than in the mother's Fallopian tube.

FIRST TAKE A CELL

★ The starting point for preimplantation diagnosis is a single cell taken from a three-day-old embryo. At this stage the embryo is still nothing more than a tiny ball of undifferentiated cells, and it is able to cope quite happily with the loss of one single cell. The removed cell can then be used to test for a whole range of genetic defects.

MORAL DEBATE

* If the genetic recipe of the cell is found to carry no genetic defects, then the embryo can be implanted into the mother's womb to continue its growth and development. If the genetic recipe reveals bad news, then the embryo will be discarded.

* The technique is morally unacceptable to some people, and in some countries it has even been outlawed.

FISHing is a new development in genetic screening

* In Britain, preimplantation diagnosis is often recommended for women who have already had several abortions because their embryos tested positive for a serious genetic disease. If there are medical risks associated with subsequent abortions, or if the woman is simply fatigued by the emotional trauma of repeated abortions, preimplantation diagnosis offers a solution.

FISHING FOR GENES

One of the latest ways of screening for defective genes is to use a technique known as **FISH (fluorescent in situ hybridization)**. Once a defective gene has been cloned, it is labeled with a fluorescent dye to make a DNA **probe**. After the probe has been added to the DNA of the cell being screened, a brief burst of heat makes all the DNA single-stranded. If the probe DNA sequence matches up with a complementary sequence in the genetic recipe of the cell, it will bind to it and show up brightly when viewed under a fluorescent microscope.

GROWTH FACTORIES

★ Biotechnology is big business. Agricultural, pharmaceutical, and medical companies are cashing in on the tools of the genetic revolution, in order to engineer new life forms with unique combinations of genes designed to suit human needs.

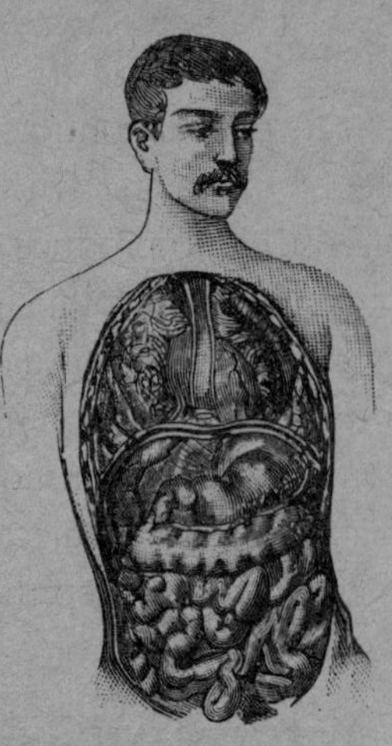

insulin is produced in the pancreas

the insulin factories

Humulin

The first genetically engineered human insulin was marketed by the Eli Lilley Corporation under the trade name Humulin.

Hormones

A hormone is a substance, manufactured and secreted into the bloodstream by a gland, that regulates the functioning of another tissue or organ somewhere else in the body. Some, but not all, hormones are proteins.

SWEET AND SOUR

★ The first commercial applications of GENETIC ENGINEERING relied exclusively on bacteria, since the techniques of gene transfer in these organisms had already been well established. One of genetic engineering's earliest successes was in producing HUMAN INSULIN. Insulin is a protein hormone, produced in the pancreas, that controls the level of sugar in the blood. People who suffer from diabetes are unable to produce sufficient quantities of insulin. Lack of insulin means that blood sugar rises to dangerously high levels, leading to a variety of complications. Consequently, many diabetics have to inject the hormone into their bodies to keep their blood-sugar levels under control.

CUT AND PASTE

* This injected insulin used to come from the pancreatic juices of cows. The only problem was that some diabetics had an allergic reaction to cow insulin. To get around this, SCIENTISTS IDENTIFIED THE HUMAN INSULIN GENE AND STUCK IT INTO THE GENETIC RECIPE OF BACTERIA. As the bacteria grew, they produced lots of the human protein, which could then be easily extracted. In effect, the bacteria were simply being used as growth factories for the production of human insulin.

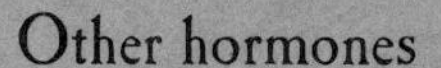

Other hormones

A variety of other human hormones and proteins are now produced in a similar way to insulin. Hemophiliacs, for example, lack a normal working copy of the gene coding for Factor VIII, the protein that enables the blood to clot properly. But provided they receive regular injections of Factor VIII, hemophiliacs can live perfectly normal lives. Today, much of the commercially available Factor VIII comes from bacteria engineered with a working copy of the human Factor VIII gene.

frost can
attack plants

Diet and dollars

Americans spend $12 billion a year on dieting aids. Yet the U.S. has the highest percentage of overweight people in the developed world. About 35 million people are putting their health at risk through overeating, and the estimated health costs amount to around $40 billion a year!

new, improved
and ice-free

THE GENE TEAM

★ Your crops are being ravaged by frost. Or you've got a weight problem. Who do you call? The Gene Team! Genetic engineers and their faithful bacterial buddies turn up in the most unlikely places, and are always ready to lend a helping hand.

DAMAGE LIMITATION

★ In temperate parts of the world, one of the biggest threats to crops and crop production is frost. Frost damage is caused not so much by low temperatures as by the formation of ice crystals. When ice forms, it expands and destroys the plant tissues.

★ Ice normally forms at 32°F (0°C). But to do so it needs a "seed"—some microscopic particle around which the ice can crystallize. Without a suitable seed, water will cool to as low as 17°F (-8°C) before freezing. So by removing the "seed," you can increase the plant's chances of surviving in cold weather.

★ The "seed" that promotes ice-crystal formation on plants is a protein known as ICE-NUCLEATION FACTOR, which projects from the cell wall of *PSEUDOMONAS SYRINGAE*, a bacterium that spends its time wandering in and around plant tissues. So to solve the problem of frost damage, scientists engineered a special strain of the bacterium with a defective version of the ice-nucleation-factor gene. ***When these "ice-free" strains are sprayed onto crops in large quantities they replace the original strain and enable the plants to survive much lower temperatures.***

fat is a leptin issue

SLIM FAST

* The dieting industry is also keeping a close eye on developments in genetic engineering, especially after the recent discovery of an "OBESITY GENE" in mice. This gene codes for a protein hormone called LEPTIN, which controls both appetite (by sending signals to the brain) and the rate at which fat is burned up. It was found that mice that carried two defective versions of the leptin gene were three times heavier than normal mice. But when injected with leptin, they soon began to lose weight.

* ***Humans also have a leptin gene*** which scientists have already cloned; clinical trials are currently underway in anticipation of its possible use as a dieting aid.

Leptin

Even if leptin does become commercially available as a diet pill, it is unlikely to be a cure-all for obesity. Indeed, levels of leptin seem to be relatively normal in obese people. This doesn't mean that additional leptin will have no impact, but it does show that obesity is affected by other genetic and environmental factors.

before after

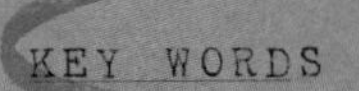

CELLULOSE:
the "woody" chemical constituent of plants' cell walls
GENE GUN:
a device for firing particles carrying DNA into a cell
REPORTER GENE:
an easily detectable gene used to diagnose whether foreign DNA has been incorporated into a cell

GENETIC IMPLANTS

★ The list of genetically engineered products continues to grow. Genetically engineered food is already sitting on the supermarket shelves. One such product is genetically engineered soy, which is used in a wide variety of foodstuffs, including breads, cakes, and margarines. Superficially, the modified soy looks no different from ordinary soy; but lurking in its genetic recipe is a bacterial gene that makes the soybean plant resistant to herbicides.

WEEDKILLERS

★ Weedkillers are an integral part of modern farming methods. Despite the huge amount of money farmers spend on trying to eradicate weeds, it's estimated that 10% of all crops are lost to the overwhelming influence of these rogue plants. One of the problems is that many weedkillers are unable to discriminate between the weeds and the crops themselves, so farmers are restricted to spraying their fields with them at certain times of the year.

★ When a strain of bacteria was discovered that was resistant

reporter genes communicate success or failure

to weedkiller, genetic engineering offered an instant solution to the problem. ***The bacterial gene for resistance was cloned and inserted into the genetic recipe of soy, producing weedkiller-resistant plants***. So now farmers can spray fields of soybean plants with weedkiller all year round, without risk of killing the crop.

FIRE POWER

★ In practice, getting foreign genes into plants and animals is trickier than getting foreign genes into bacteria. Plant cells, with their thick walls of cellulose, are obstinate recipients. One way around the problem is to fire microscopic DNA-coated gold bullets from a GENE GUN through the cell walls.

LET THERE BE LIGHT

How do scientists know whether a foreign gene has been successfully incorporated into an organism's genetic recipe? An easily detectable **"reporter gene"** is attached to the foreign DNA. One of the most cunning reporter genes codes for **luciferase**, an enzyme that emits light and makes fireflies glow in the dark. To find which cells have incorporated the foreign DNA, scientists simply look for the ones that are shining brightly.

SUICIDAL POTATOES AND GAS-FREE BEANS

* There seems to be no limit to the scientific imagination of food engineers. Hundreds of new types of food are currently being tried and tested, and many of them may soon be finding their way to a supermarket near you.

no more nasty odors

Bettering Beans

Britain imports more than 400 million cans of Heinz baked beans each year. This figure would be even higher if it weren't for the sizeable proportion of the population who are won't eat beans because of their flatulence-inducing potential. But their worries may soon disappear now that a Cambridge scientist has engineered a new bean which lacks the enzyme responsible for the sudden and unannounced "gas attacks."

SELF-DESTRUCT

* One of the most bizarre examples of a new food variety is a ***potato that has been genetically engineered to commit suicide***. This isn't quite as ludicrous as it sounds. The potato is extremely selective about when it kills itself. Only when infected with a fungus does it decide enough is enough.

* The enzyme that kills the potato is called BARNASE, a naturally occurring toxin found in a species of soil-dwelling bacteria. Scientists cloned the barnase gene and then inserted it into the potato's own genetic recipe. The gene has been engineered so that it is only switched on when the potato is under fungal attack.

* The scientific reasoning behind the new potato variety is subtle and clever. Fungal diseases, such as potato blight, can decimate potato crops. Normally these

diseases can only be controlled by the generous application of fungicides. But the new potato could do away with the need for fungicides altogether. As soon as a potato is infected, it commits hara-kiri. This is bad news for the individual potato, but good news for potatoes in the rest of the field. ***The potato's self-sacrifice prevents the disease from spreading.***

Cosmetic changes

Some genetic changes are more cosmetic. Genetic engineering has been used to artificially enhance the color and shape of fruits and vegetables. A brand of engineered tomatoes is now available that is resistant to bruising, thereby prolonging its shelf life. But most strange of all, scientists are currently looking at ways of altering the familiar shapes of fruit and vegetables altogether, in order to make them easier and more economical to pack. So it may not be long before we see the square apple and straight banana in our fruit bowls.

would you like a square or round cabbage today ma'am?

GENETIC INSECTICIDE

* Many kinds of ***fruit and vegetables have been rendered resistant to insect pests by using genes for bacterial poisons that are toxic to insects but not to the plants***—nor, reassuringly, to the humans who eat them.

UNNATURAL FOOD?

★ Not everyone is happy with genetically engineered foods. Some argue that they are a violation of nature and carry potential risks to our health and the environment. Does this alarmist view simply reflect a general fear of new technology, or is there real cause for concern?

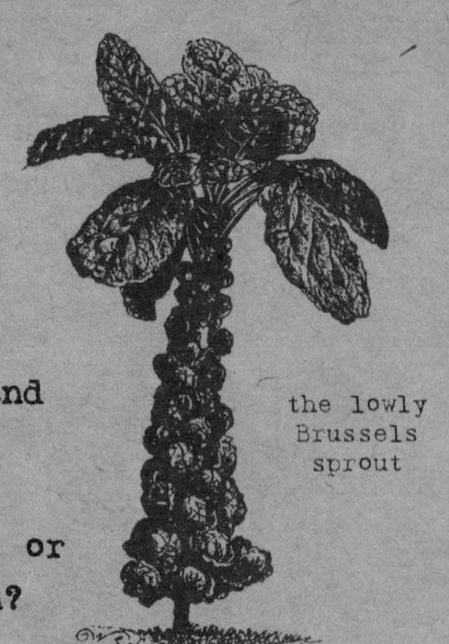

the lowly Brussels sprout

TRANSFERS

Viruses and plasmids have a particular knack for picking up genes from one species and inserting them into the genetic recipes of another. This **horizontal gene transfer** could explain why proteins such as hemoglobins are found not only in animals but also in leguminous plants!

AGE-OLD TRADITION

★ Let's face it, humans have been tampering with nature for thousands of years. Pigs, sheep, and cows, the mighty eggplant, the lowly Brussels sprout, and the humble radish—these are just some of the things we take for granted when we sit down for dinner. And none of them would exist without the long history of human interference in nature. In one sense, genetic engineering is simply a continuation of a centuries-old tradition. SELECTIVE BREEDING and genetic engineering have identical aims—to produce organisms that are adapted to the needs of humans rather than those of nature.

★ Some may argue that selective breeding of animals and plants is a more natural form of human interference than the gene-wrenching manipulation of genetic engineering. But the distinction is a blurred one: viruses regularly transfer genes

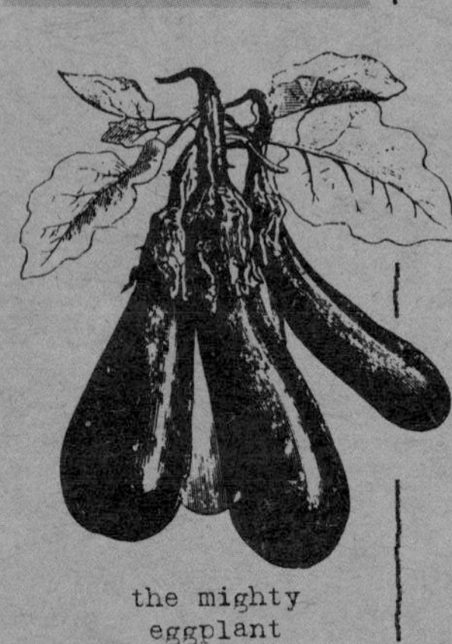

the mighty eggplant

between different species in the wild, and the success of genetic engineering often relies on this natural ability of organisms to move genes around.

genes work in teams

* What of the more serious accusation that genetically altered foods may be hazardous to our health? ***Moving a gene into a completely different genetic recipe can have unpredictable consequences, because genes often work together as a team***. Like the transfer of a star player between football teams, the transfer of a single gene does not guarantee success. There are cases where genetic manipulations have produced unwanted characteristics—such as foods with an abnormally low nutritional content, or some new allergenic chemicals. But rigorous testing should, in theory at least, ensure that any difficulties are ironed out before new types of food appear in grocery stores.

ROGUE SUPERWEEDS

Environmental worries are, perhaps, the most realistic concern. There is a theoretical risk that foreign genes could spread further than was originally intended. Take the example of the soybean plants made resistant to weedkillers. If these plants were to breed with their wild relatives, or if wild viruses were to conjure up some genetic cut-and-paste tricks, then the resistance gene could spread to other plant species, giving rise to rogue superweeds. These could monopolize the countryside and drive many rare and endangered plants to extinction.

pigs have been bred selectively for years

THE FLIP SIDE OF CYNICISM

★ Environmental pressure groups are keen to promote the idea that genetic-engineering companies are motivated more by greed than by the desire to improve the lot of the human race. But this is as much a criticism of capitalism as of genetic engineering itself.

fat cats need leptin

FOOD FIGHTS

Opposition to genetically engineered food seems to be much stronger in Europe than in the U.S. In Britain some supermarkets have removed genetically engineered products from their shelves. Even Prince Charles has spoken out against genetically engineered food.

WASTE NOT WANT NOT

★ IT'S EASY TO BE CYNICAL ABOUT GENETICALLY ENGINEERED FOOD. So here goes. Do we really need all these new products when so much food in the Western world already goes to waste? Aren't "gene foods" just one **giant marketing exercise**, of which the primary beneficiaries are the companies that manufacture them? **And haven't we already done enough damage to the environment without compounding the problems by introducing novel and unpredictable genetic creations?**

what happened to the food we used to eat?

HUNGRY FOR MORE

★ Okay, that's the cynicism over with. What's the flip side? What are the potential benefits on offer? These are many and varied. For a start, ***genetically engineered plants could actually improve the state of our chemically ravaged countryside***.

bug-resistant plants will reduce the need for pesticide

Many of the latest products have been specifically designed to have innate genetic resistance to insect and fungal pests, thereby reducing the need to spray chemical pesticides and fungicides all over the place.

★ Could genetic engineering benefit the hungry? In Bangladesh, the soil in the Ganges delta is so salty that it is often difficult to grow food crops like wheat. One solution is to engineer wheat which contains a gene from a native salt-loving plant.

THE EDIBLE VACCINE

Another ingenious development is the edible vaccine. In the developing world, diarrhea kills thousands of people every year. Although vaccines are available, there are problems of cost and distribution. But scientists hope that a genetically engineered banana will provide a more efficient way of distributing the vaccine. The banana, which could be grown locally, carries a gene for a protein vaccine conferring resistance to the bacteria and viruses that cause the diarrhea.

MEDICINAL MONEY

★ Companies that manufacture new types of food are in the business of making money. But some of the latest gene products do seem to offer real medical and agricultural advances that could have a big impact on many human lives.

OF MICE AND MEN

mice are ideal for genetic research

★ Unlike plants, most genetic engineering in animals is done in the interest of medical research rather than for culinary purposes. So those anticipating the self-cooking chicken or the boneless cow will be disappointed. Nevertheless, the applications of recombinant DNA technology in animals are diverse. Much of our scientific understanding of human development and genetic disease has come from the study of transgenic mice.

MOUSE MODELS

★ Mice are the mainstay of genetic engineering in animals. Because ***most mouse genes do pretty much the same thing in mice as human genes do in humans,*** mice are seen as ideal animals for the study of human development and disease.

★ Getting foreign genes into cells is easier in animals than it is in plants. Animal cells are much more squishy, so a gene can be micro-injected straight into a fertilized egg. It's hit or miss, but at least sometimes the foreign gene will incorporate itself into a chromosome. The egg is then implanted into the

KEY WORDS

TRANSGENIC ANIMAL: an animal containing a foreign gene (a transgene) inserted into one of its chromosomes

GENE CASSETTE: a segment of DNA inserted into a gene to prevent it from making its normal protein product

KNOCKOUT MOUSE: a mouse which has had both working copies of a gene disabled

womb of a female mouse, to continue its growth and development.

* ***Genetic manipulations in mice enable scientists to answer basic questions about the function of individual genes***. One of the best ways of understanding what genes do in mice is to look at what happens to the mouse when things go (deliberately) wrong. After a mouse gene has been cloned, the gene is deliberately inactivated by inserting a GENE CASSETTE into its DNA sequence. This disabled gene is then injected into a fertilized mouse egg to replace a normal working copy.

* Scientists can examine the effect on the mouse of having a single defective copy. By breeding two heterozygous mice together, they can create a homozygote or knockout mouse, and look at the effect of having two defective copies.

RAT AND MOUSE

The first successful transplant of a foreign gene into another animal occurred in 1982 when a rat growth-hormone gene was inserted into a fertilized mouse egg.

Mice – A Lament

When you think about it, mice have a pretty raw deal. In the wild they live pitifully short lives, in which they are forever fearful of the neighborhood owl or the local cat. Yet their laboratory cousins fare even worse—despite the warmth of their surroundings and their isolation from owls or cats.

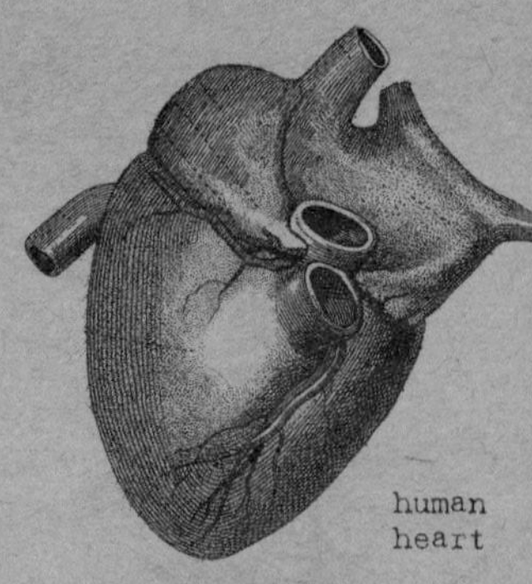

human heart

HEART TO HEART

* Human heart transplants are now regarded as fairly routine operations. But unfortunately the demand for replacement hearts far exceeds the supply: there simply aren't enough donors to go around. With a bit of genetic tweaking, however, pig hearts could do the job just as well—and there is certainly no shortage of pigs.

KEY WORDS

XENO TRANSPLANTATION: the transplant of an organ from one species into another species (from the Greek *xenos,* meaning "strange" or "foreign")

THIS LITTLE PIGGY...

* Pig hearts and human hearts both do the same job—pumping blood around the body. They're also remarkably similar in size and shape. But you can't just take a pig's heart and stick it straight into a human being. The body will recognize the heart as foreign and REJECT it. Proteins on the surface of the pig's heart cells are what give the game away.

* Genetic engineers have gotten around this problem by inserting the genes that make human

I give you my heart

caught red-handed

cell-surface proteins into a pig's genetic recipe. ***so the engineered pig expresses human proteins on the surface of its heart cells.*** If the pig's heart were then transplanted into a human, the body would be tricked into thinking that it was a human heart.

HIDDEN DANGERS

* Nobody has attempted to transplant a pig's heart into a human being yet. This is not because of fears that the pig heart won't work properly—it almost certainly would—but because of the more sinister threat of transferring pig viruses and diseases to humans. AIDS and "mad-cow disease" have made us only too aware of the way in which diseases, given half a chance, will happily skip from one species to another, with severe consequences.

THE IMMUNE SYSTEM

The immune system is the body's own molecular defense against foreign invaders. If it didn't exist, then we'd soon be taken over by a whole host of infectious agents and would die. The immune system is made up of a variety of cells and proteins that patrol the body looking for **antigens** – molecules which they do not recognize as "self." Antibody proteins recognize and bind on to these antigens, triggering a cascade of responses designed to eliminate the intruder.

GENE THERAPY

ethical and moral arguments continue to rage

* The only way to cure a human genetic disease is to replace the defective gene, or genes, responsible for the disease with normal working copies. Gene-replacement therapy is now being developed for use in the treatment of some genetic diseases. But ethical and moral considerations preclude the use of gene therapy on human embryos.

MAN OR MOUSE

* Altering the genetic recipe of a pig or mouse embryo is one thing—but tampering with the genes of a human embryo raises all sorts of difficult ethical questions. Though the technology already exists to manufacture new genetic recipes in very early human embryos, research into so-called GERM-LINE GENE THERAPY has been banned. *Once a new gene has been inserted into a fertilized egg, it will form part of the genetic recipe of every cell in the body, including the future germ-line cells*. This is too close to the image of "designer babies" and eugenics for most people.

should you be allowed to design your baby?

TARGETING TISSUES

* So, instead, scientists are looking at ways of getting genes into the cells of tissues that are most affected by a disease. Because cystic fibrosis primarily affects the lungs, and the lungs are exposed and fairly easy to get at, the disease has been seen as a promising candidate for gene therapy.

TRIAL THERAPY

The first trials into gene therapy were begun in 1990 by French Anderson of the University of Southern California. He treated children suffering from Severe Combined Immune Deficiency (SCID).

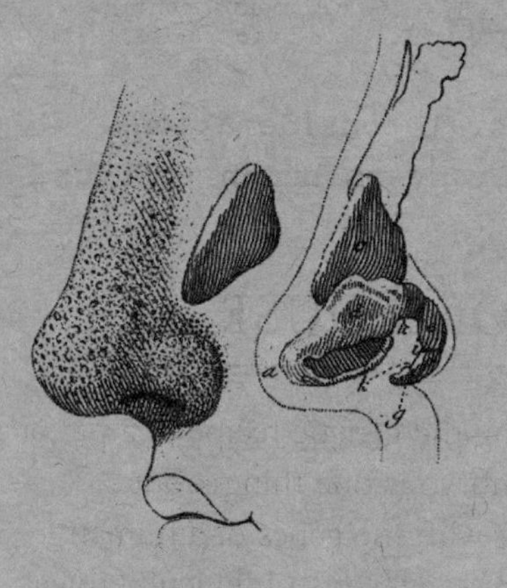

VIRAL RESCUE

* A virus is used to get the normal working copy of the cystic fibrosis gene into the affected cells. Viruses normally cause disease when they get inside their host—so before it can be used as a vector, the virus has to be genetically altered to make it more benign. Once the cloned cystic fibrosis gene has been inserted into the disarmed virus, the virus is then inserted through the nose and into the lungs of the patient. During infection the virus inserts its own DNA, carrying the normal copy of the human gene, into the DNA of the affected lung cells.

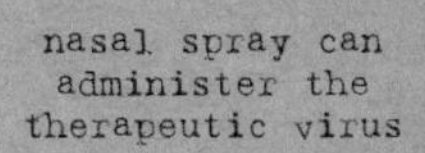

nasal spray can administer the therapeutic virus

Nasal therapy

There have already been some relatively successful trials of gene therapy for cystic fibrosis sufferers—and it is possible that in the future the virus may be administered by means of a simple nasal spray, thus providing a quick, easy and painless cure.

GENETIC FINGERPRINTS

Imagine a genetic recipe as 46 very long pieces of string, each piece of string representing the DNA molecule that makes up one of the 46 chromosomes. A restriction enzyme works like a pair of molecular scissors. Wherever it recognizes a specific sequence of DNA letters, it cuts the string. Because different people have different DNA sequences, the places where the strings are cut are different for each individual. So after the strings have been chopped up, each person has their own unique combination of string fragments, which vary in size and number.

everyone has a unique fingerprint

DNA FINGERPRINTING

* Genes are a part of what defines our identity. Their visible signatures, in the form of fingerprints or blood types, have long been used to identify individuals in criminal investigations and paternity disputes. But in addition to these traditional methods of fingerprinting and blood typing, DNA technology has provided some powerful new tools for sniffing out suspects. DNA fingerprinting has made it much more difficult for criminals to, literally, get away with murder.

THE ONE AND ONLY

* Fingerprints have been used for identification purposes since the early 1800s. ***Everyone has his own unique set of fingerprints***. Even identical twins, who share an identical set of genes, display slight differences in the fine pattern of skin ridges that make up their fingerprints. So differences in fingerprint patterns are due not only to genes, but to environmental factors during fetal development.

* Fingerprints are an accurate way of matching a suspect to the scene of a crime. Unfortunately, criminals are only too aware of this, so they wear gloves to conceal their tell-tale marks of identity. But short of

turning up in a plastic bubble, they cannot avoid leaving behind fragments of skin, hair, and other bodily tissues—all of which contain copies of their unique genetic recipe.

only a plastic bubble would not leave any trace

* DNA FINGERPRINTING was first developed in the 1980s by **Alec Jeffries** at the University of Leicester, in England. What makes the technique so powerful is its sensitivity. A tiny speck of tissue provides sufficient copies of a genetic recipe to obtain a genetic fingerprint. This is why the technique has been so widely adopted by forensic scientists.
* After the genetic recipe has been extracted from a tissue sample, restriction enzymes are used to cut it up into smaller fragments which can be separated from one another by gel electrophoresis.

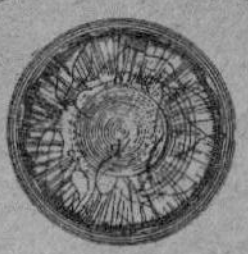

retinal scans can also be used as identity checks

Eye–dentity

Scientists recently discovered that each of us has a unique pattern of blood vessels on the retinas of our eyes. ***Retinal scans*** *are therefore now being used by the security conscious as additional proof of an individual's identity.*

Technology

Although the technology underlying genetic fingerprinting is state-of-the-art, the principle behind its use differs little from traditional fingerprinting. A DNA fingerprint is a mark of a person's identity—but one based on their unique DNA sequence, rather than the unique pattern of ridges on their fingertips.

get the DNA onto paper

KEY WORDS

GENOME: all the DNA in the genetic recipe of an organism

SOUTHERN BLOTTING: a method of transferring DNA fragments from a gel onto a piece of nylon paper, and then visualizing the fragments through **hybridization** with a DNA probe. The technique's name has nothing to do with compass bearings—it was named after its inventor, **Ed Southern**.

AUTORADIOGRAPH: an image of the distribution of radioactive particles produced on a radiation-sensitive film

PICKING OUT THE PRINT

* Even after the DNA fragments have been separated on a gel, they will still be invisible. The DNA can be stained with ethidium bromide, but there won't be enough of it to show up under ultraviolet light. So another method, known as "Southern blotting," is needed to visualize the chopped-up DNA fragments.

SOUTHERN BLOTTING

* First, the DNA fragments have to be transferred from the gel onto a piece of nylon paper. This is laid on the top of the gel and, under pressure, the DNA fragments are forced through the gel onto the paper, where they stick fast to the long polymeric molecules of the nylon. Moving the DNA from gel to paper doesn't change the relative positions of the fragments to one another.

UNZIPPING

* The next stage is to incubate the paper at a high temperature. This unzips each of the double-stranded DNA fragments into their two complementary strands. Again, the locations of the different sized fragments on the paper remains unchanged.

PROBING

* Then comes a really nifty bit of molecular trickery. A solution containing a DNA PROBE is added to the paper. The probe is a bit of artificially made single-stranded DNA sequence, usually about a few hundred DNA letters long, that has been tagged with a radioactive isotope.

finding the DNA sequence is as easy as ABC

* In DNA fingerprinting, the DNA probe has the same sequence as one of the junk DNA sequences found in multiple copies in all human genetic recipes. When the single-stranded DNA probe is added to the nylon paper, it will bind to any DNA fragments that contain a complementary sequence. Since the junk DNA sequence is ubiquitous throughout the human genome, the probe will bind to a large number of fragments.

* Finally, the locations of the DNA fragments that bound to the radioactive probe are revealed by overlaying the nylon paper with a piece of radiation-sensitive film. When the film is developed then, at last, we have a visual picture of the DNA fragments.

hybrids are interbred

Hybridization

In biology, hybridization has two meanings. It can refer to interbreeding between two different species or varieties, or it can mean complementary base pairing between two DNA strands from different sources (i.e. a DNA probe and a DNA fragment).

your DNA gave it away

chance plays a part in DNA fingerprinting

THE NECESSITY OF CHANCE

* Once a DNA fingerprint has been obtained, the next job is to find a suspect whose own DNA fingerprint matches the one found at the scene of the crime. Because the chance of two different people sharing the same DNA fingerprint is extremely low, a profile match can be presented as fairly damning evidence in court.

REASONABLE DOUBT

* If the DNA fingerprint of a suspect does not match the one obtained from the tissue samples found at the scene of a crime, then the two DNA samples must have come from two different people, so the suspect is exonerated. But what happens if the DNA fingerprints do match up? Is this conclusive proof of guilt?

* Securing a conviction based on a matching DNA fingerprint rests on the assumption that fingerprints vary so much between different people that the chance of two people sharing the same fingerprint is low. But although everyone has their own unique genetic recipe, DNA fingerprinting only samples a proportion of that recipe. ***There may be a chance, however slim, that two people do in fact share the same DNA fingerprint as revealed by genetic testing.***

SUSPECT DNA

DNA data banks have been established in the U.S. and Britain. Screening the catalog of fingerprints is the first step in tracking down possible suspects.

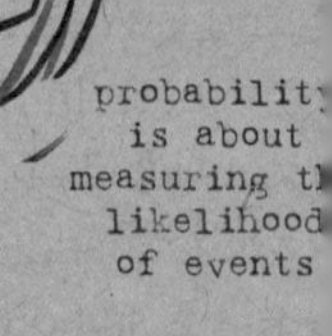

probability is about measuring the likelihood of events

guilty! (or innocent?)

* So scientists have to work out the probability of a chance match. Despite the fact that the DNA fingerprint from the scene of the crime matches that of the suspect, the scientist has to work out what the chances are that it came from someone else.

PROBABILITY

* The estimate of PROBABILITY has to be based on a sample population of which the suspect is a member. This is important, because ***two people from the same ethnic group are more likely to share genes, and therefore DNA fingerprints, than two people drawn at random from the population as a whole***. However, working out the probability of chance matching is never an exact science, and lawyers can exploit this element of doubt in court. In Britain some people have been convicted on DNA evidence alone—but without additional incriminating evidence, this is still quite rare.

Incompetence

There are more prosaic reasons why DNA evidence may not stand up in court. Any suspicion of incompetence in the way tissue samples were collected from the crime scene, or during the laboratory testing itself, can lead to allegations that the samples were contaminated by DNA from other sources.

do I have to repeat myself?

REPEAT AFTER ME...

* Other forensic DNA technologies have emerged since the introduction of DNA fingerprinting. PCR techniques have been incorporated into the forensic scientist's tool box, to uncover the repetitive DNA sequences that are scattered throughout our genetic recipes.

FRAGILE-X SYNDROME

Fragile-X syndrome is a relatively common congenital mental disorder in humans caused by a tri-nucleotide repeat sequence stuck in the middle of a gene on the X chromosome. How the sequence got there is a mystery, though it may have been part of a jumping gene. The repeat sequence completely disrupts the gene's function. With each generation the number of repeats increases—with the result that children always have worse symptoms than their parents.

TANDEM REPEATS

* One of the problems for forensic scientists is that unless the tissue samples they collect from crime scenes are relatively fresh, the genetic recipe within the cells of the tissue will have become degraded. In other words, microbes will have begun chopping it up. For obvious reasons, this can make conventional DNA fingerprinting impossible.

* But ***PCR has the advantage that it only requires a small section of intact DNA***, rather than the genetic recipe in its entirety. It will work provided the DNA has not been chopped slap-bang in the middle of the sequence you want to amplify.

speed is vital at the crime scene

* One of the most common applications of PCR in forensic science is in the study of VNTRs (Variable Number of Tandem Repeats). These are noncoding

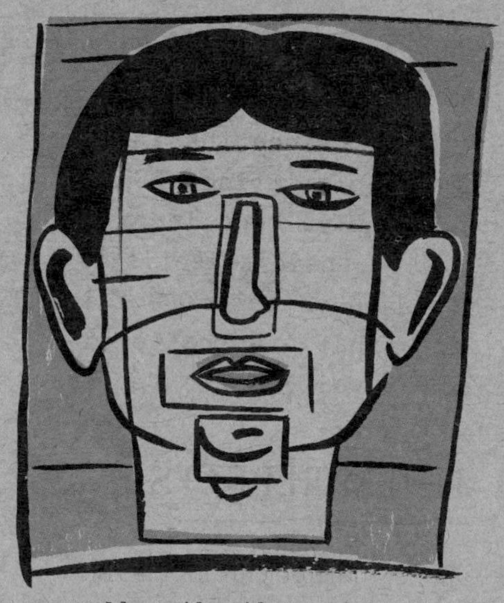

could we identify a criminal from a genetic profile?

sequences, found throughout the genome, made up of endless "tandem repeats"—for example, ATATATATATATAT. The number of repeats varies enormously between individuals. When these regions are amplified by PCR, different people produce tandem repeats of different lengths, which can be separated by size using gel electrophoresis. And because PCR produces so many copies of the amplified sequence, there is no need to fool around with the long-winded complexities of Southern blotting. The DNA can be visualized by staining it with ethidium bromide and viewing it under ultraviolet light. ***By looking at many of these individual tandem repeat sequences, scientists can produce a DNA profile that is similar to a DNA fingerprint.***

TWOS, THREES, FOURS . . .

Repetitive sequences don't just come in repeats of twos. Threes, fours, and much larger repeat units are well known. All these repetitive sequences have much higher mutation rates than coding genes, and changes involve the addition or deletion of one or more of the repeated units.

If the Face Fits...

When trying to solve a crime, police forces often publish an artist's impression or a computer-originated composite picture of a suspect, based on the recollections of eyewitnesses. In the future, a genetic recipe may be all that is required to build a picture of a criminal. Scientists in the UK are currently constructing a database containing the genetic recipes of a group of volunteers plus detailed 3-D images of their faces, in order to see how well DNA can predict facial features.

sitting on the tree of descent

CLUES TO THE PAST

* Evolutionary change ultimately comes about through the steady accumulation of new mutations in DNA. By comparing the DNA sequences of different species, it's possible to trace their evolutionary history. DNA sequencing has even offered insights into our own biological past.

ONLY CONNECT

Taxonomy is the study of the classification of living things. Gone are the days when taxonomists spent endless lonely hours measuring the lengths of bones. Nowadays, they spend equally endless lonely hours sequencing genes from different species to work out how they are related to one another.

SPLENDID ISOLATION

* Imagine the following hypothetical evolutionary scenario. A population of aardvarks is happily minding its own business when a splinter group suddenly decides to head off for greener pastures. Each individual in the splinter group will take with it the aardvark genetic recipe. But as the years pass, genetic changes will begin to accumulate in the two

excuse me, could you tell me the way to my evolutionary history

isolated populations. The extent of this divergence will be correlated with the length of time they have lived apart.

* The aardvarks illustrate the general principle that ***the longer the interval since two populations or species shared a common ancestor, the greater is the difference between their DNA sequences***. So, by reversing the logic, we can use the DNA sequence differences between species to work out the evolutionary relationships between them.

VARIABLE RATES

* The choice of which genes to look at is an important one. ***Different regions of DNA evolve at different rates***. Noncoding sequences, for example, evolve much more rapidly than protein-coding genes, because mutations can steadily accumulate in the genetic garbage without any detrimental effects on the organism.

* Slowly evolving genes are not very informative for investigating the evolutionary relationships of closely related species, because they will have accumulated few, if any, differences. In general, fast-evolving genes are used to distinguish closely related species; and slowly evolving genes are used for distantly related species.

Close relatives

DNA sequencing has shown that chimpanzees are our closest living evolutionary relatives. DNA sequences of humans and chimps differ by only 2.5%. Given that the genetic recipes of humans differ from one another by an average of about 1%, this means that our own genetic recipes differ from those of chimps by about 20 DNA letters in every thousand.

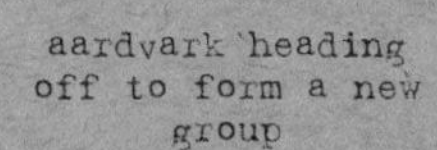

aardvark heading off to form a new group

Africa may be the cradle of human history

DETECTING DIFFERENCES

DNA differences between human populations are slight. Mitochondrial DNA is often used to distinguish human populations, because it has some noncoding genes that evolve at a very high rate, making differences relatively easy to detect.

the genetic history of African peoples has been studied

GEOGRAPHICAL GENES

*** Tracing the ancestry of families, populations, and nations is a popular pastime among historians and archaeologists. Cultural artifacts and other human remains contain clues to our past, and their geographical distribution has illuminated patterns of human ancestry and migration. But genes contain much more reliable clues, and DNA technology is adding an extra dimension to studies of human history.**

OUT OF AFRICA

* Before DNA technology, the only way of investigating early human origins was through their fossil remains. Because these are difficult to date accurately, there has always been ample scope for speculation and conflict over when and where humans first originated. One theory suggests that *Homo sapiens* evolved simultaneously in Africa, Europe, and Asia. An alternative "out of Africa" theory postulates that humans had an exclusively African origin and subsequently spread to other parts of the world. Geneticists go for the latter theory.

* A number of recent genetic studies have found that DNA differences between modern African populations are much greater than those between populations

African populations may be the oldest in the world

from other parts of the world. That more DNA differences have accumulated within African populations implies that ***African populations are older than other human populations***.

MAKING A DATE

* DNA data provides support for the "out of Africa" hypothesis. But geneticists have been able to go further with this information and actually put an estimate on the time when humans first evolved. This calculation rests on the assumption that mutations accumulate in the DNA at a steady rate, like a ticking clock. By counting the number of accumulated differences and knowing the rate at which mutations occur, you can extrapolate backward to pinpoint a date of origin. Using this technique, it has been estimated that ***humans originated about 100,000 years ago***.

Genetic baggage

When some human populations decided to pack their bags and leave Africa, they didn't just take their suitcases, language, and culture; they took copies of their genes as well. Like languages and cultures, the present day distribution of genes records patterns of human migration. The geographical patterns of genes in modern humans suggest that Europeans wandered into Europe via the Middle East about 45,000 years ago. Native American people seem to derive from two separate migrations, both from Asia, around 10,000 and 30,000 years ago.

the werewolf condition may be a genetic disorder

MIGRATION AND MADNESS

* The history and movement of populations leaves its mark on the present day distribution of genes. The global distribution of the dominant genetic disease porphyria variegata provides one of the most striking examples of this phenomenon.

HEADING SOUTH

* In most parts of the world, porphyria variegata is very rare. But among the Afrikaans-speaking people of South Africa, it is much more common. Today, about one out of every 300 Afrikaners has the disease. Afrikaners are descendants of a Dutch couple who married and founded a settlement in South Africa in the late 17th century. Although porphyria is rare in Holland, by chance either one or both of the couple must have had the porphyria gene in their genetic recipes.

King George III may have suffered from porphyria

Werewolves and Kings

The legend of the werewolf may have originated with the disease porphyria variegata. The effects of the disease are very variable, but in extreme cases sufferers can experience such diverse symptoms as extreme sensitivity to light, excessive hair growth on exposed skin, blood-red urine, and bouts of madness. It has been suggested that the disease was responsible for the "madness" of the English King, George III.

INBRED DISEASES

* The Dutch couple produced a large family, so the gene would have been passed on to some of their children and grandchildren, and so on. Although the settlement grew rapidly, language and religious barriers meant that the population was effectively cut off from the outside world and was therefore exposed to the dangers of INBREEDING.

porphyria was brought to South Africa by the first Dutch settlers

* The genetic risks of inbreeding provide a good biological reason why incest has become a cultural taboo in most human societies. Close relatives share more genes in common with one another, including the genes which cause disease—so if you marry a close relative and have children, there is a much greater chance that you will produce a child with two copies of a disease-causing gene.

REMOTE RISKS

Wherever communities remain small and are isolated by religious, cultural, or geographical barriers, there is a much greater risk of a genetic defect spreading in a population through inbreeding. For example, a recessive genetic disease called **BIDS syndrome**, which causes extremely brittle hair and intellectual impairment, is relatively common in the religiously isolated Amish community of North America, but unknown elsewhere.

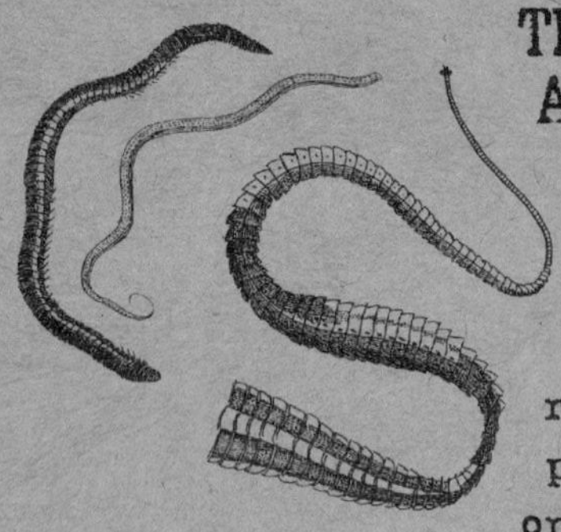

DNA from worms millions of years old has been retrieved

THE ANCIENT AND THE MODERN

* The death of an organism invariably spells disaster for its DNA. The DNA is broken down along with the rest of the cellular bits and pieces, as a host of micro-organisms gorge themselves on the banquet of molecules. But occasionally there are some remarkable acts of preservation.

BACK IN TIME

* Mitochondrial DNA (mtDNA) is generally more resistant to degradation than NUCLEAR DNA and can remain intact for a few thousand years. Most impressive of all, however, are the rare instances when an entire organism and some of its DNA have been preserved for millions of years. MtDNA can be obtained from museum specimens of long-extinct animals. It has also been extracted from long-dead mammoths found frozen and preserved in Siberian ice.

ancient DNA gives clues to evolutionary history

Jurassic Park

Many of the insects found preserved in amber existed at the same time as the dinosaurs. In Michael Crichton's Jurassic Park, the starting point for the genetic regeneration of the dinosaurs is a mosquito that was immersed in amber just after dining on a meal of dinosaur blood. Outside Hollywood, this is implausible. Although the blood might contain the odd dinosaur gene, most of the DNA would be degraded. But if someone should ever encounter an entire chunk of dinosaur flesh preserved in amber...

DNA can be taken from the remains of extinct animals

* Ancient mtDNA can be used to work out evolutionary relationships between extinct animals and their modern living relatives. It is also useful in calibrating molecular clocks. The farther back in time you go, the more accurate your estimate should be of the rate at which mutations accumulate in the mtDNA.

RESIN RESIDENTS

* The most impressive preservation stories are those involving amber, a hardened tree resin. Insects, snails, worms, and lizards have all been found in amber, looking as fresh as the day they died, despite the fact that some of them are over 100 million years old. Intact DNA has been recovered from some of the insect specimens and sequenced—and some ***30-million-year-old bacteria have even been revived***. When times get tough, bacteria form spores. In effect they shut down all metabolic activity and wait for environmental conditions to improve.

DNA from a Mummy

DNA has been extracted from a 2,400-year-old Egyptian mummy. The DNA was badly degraded, but it was possible to sequence several thousand base pairs. No recognizable genes were identified, but most of the human genome is junk DNA, so finding a gene in a few intact remnants is unlikely.

BUZZ OFF

Sleepy spores

Bacterial spores were recovered from the gut of an amber-preserved bee. When they were provided with nutrients, the bacteria suddenly woke up and emerged from their shells! Despite being separated by 30 million years, DNA sequencing revealed that the ancient bacteria were similar to present-day bee bacteria.

CANCER

★ If there's one disease that sparks more fear and dread in our minds than all others, it's cancer. This fear is not an irrational one. The harsh reality is that one in four of us will die of some form of cancer. Cancers are caused by mutations in genes that control cell division.

there are dozens of books giving the findings of cancer research

Cancer causes

It is estimated that about 10 percent of all cancers are directly caused by the inheritance of faulty genes. But the majority result from an interaction between environmental factors—our diet and lifestyle—and the genes we inherit.

SENDING THE WRONG SIGNALS

★ Cell division is required for organisms to grow. But it's important that their division is closely monitored and doesn't get out of hand. It's controlled by two different types of genes that have opposite effects. ONCOGENES promote cell division, whereas ANTI-ONCOGENES suppress it. The interaction of these two types of gene ensures that our organs and tissues grow to their right size. By the time we reach our early twenties, most of our cells have stopped dividing and we are "fully grown."

★ But mutations in the oncogenes and anti-oncogenes of a cell can get the signaling system all messed up. A cell which had stopped dividing and was happily minding its own business is now

one in four of us will die of cancer

transformed into a cancer cell that starts dividing again. As the cancer cells divide, they begin to form a tumor, which can disrupt the normal function of the tissue from which it originated.

★ If at this stage the tumor is removed by a surgeon, then there is still a good chance of a full recovery. But if the tumor grows even larger, some of the cancer cells bud off and colonize other tissues to form secondary tumors. By this time, the chances of recovery are slim.

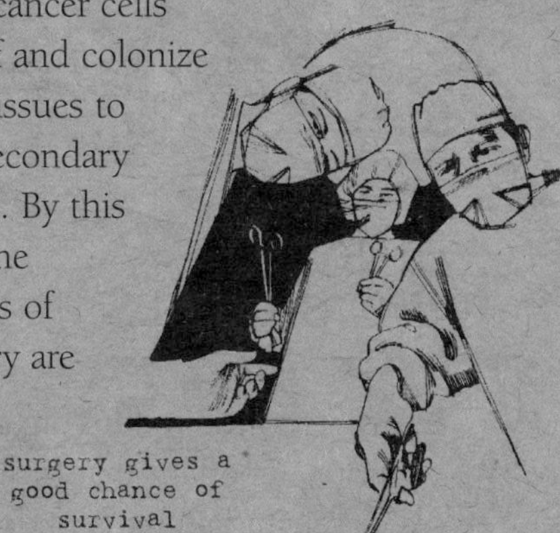

surgery gives a good chance of survival

RISK TAKING

Cancer-causing mutations occur in the same way as mutations in other genes. The reason that smoking causes lung cancer isn't because the carcinogens in cigarette smoke specifically target the oncogenes and anti-oncogenes. It's simply that they increase the overall mutation rate throughout your DNA, which therefore increases the chances of a mutation in a cancer-causing gene.

INHERITANCE

★ People do vary in their susceptibility to certain cancers. If we all smoked 20 cigarettes a day, the risk of developing lung cancer would not be the same for everyone. Some people may inherit a faulty version of an anti-oncogene, which will become part of the genetic recipe of every cell in the body. It will then only take one mutation in the other working copy for the problems to begin. If you inherit two normal working copies of the anti-oncogene, then the risk is much lower—since two mutations will then be required.

Natural sunscreen

There are other ways that genes can affect our chances of getting cancer. White people, particularly those who live in the sunnier parts of the world, develop skin cancer more frequently than Blacks. This is not because Whites are more likely to inherit faulty genes, but because Black people have more melanin in their skin, which protects their DNA from the damaging effects of ultraviolet light.

the war against cancer continues

CANCER WARS

* The war on cancer has already cost billions of dollars, and we still have very little to show for it. Most of the major medical breakthroughs have been in the detection and diagnosis of cancers, rather than actual cures. But new developments in genetic engineering and gene therapy offer hope for the future.

SCRAWNY CELLS

The effective treatment of cancers relies on a combination of early diagnosis, surgery to remove the tumor, radiotherapy, and chemotherapy. Radiotherapy involves firing radiation at the area of infection to kill any remaining cancerous cells. Despite their potential to divide, cancer cells are very scrawny, and chemotherapy exploits this weakness through the use of toxic drugs, which kill the cancer cells, but leave the normal cells relatively unharmed.

NEW DEVELOPMENTS

* Many cancers are nipped in the bud by the body's own immune defense system before the tumor has had time to get a stranglehold. White blood cells are sometimes able to infiltrate small cancers and release an enzyme called TUMOR NECROSIS FACTOR (wisely abbreviated to TNF), which zaps the tumor cells (see diagram and text opposite).

FAT CHANCE

* An alternative approach uses more subtle tactics, and relies on the immune system launching a more concerted and organized attack on the offending cancer cells. Whenever an infectious agent, such as a virus or a bacterium, gets inside your body, their cell-surface proteins are recognized as foreign by the body.

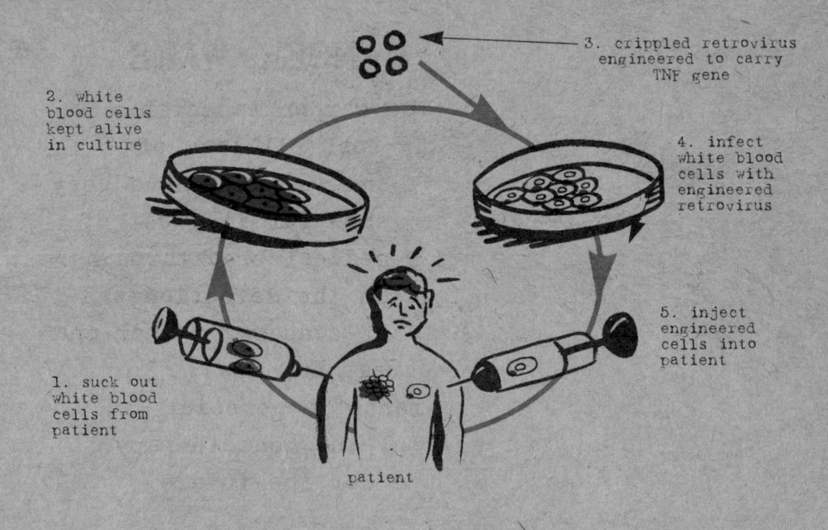

GENE THERAPY FOR TNF GENE

Although white blood cells are unable to contend with the bullying tactics of larger tumors, genetic engineering could give them more of a fighting chance. **One possibility is to insert multiple additional copies of the TNF gene into each white blood cell, to boost their cancer-beating performance**. Having extracted some white blood cells from the patient's blood, cloned TNF genes could then be inserted into their genetic recipes. The fortified cells would then be injected back into the patient.

The immune system then goes on a full offensive to quash the intruders.

* This system usually doesn't work against cancer cells, for the simple reason that cancer cells are the body's own cells and so don't present any foreign signals. ***But it could work if foreign genes coding for alien cell-surface proteins were inserted into the DNA of the cancer cells.*** One way of doing this is to use liposomes—hollow globules of fat.

The liposomes containing the foreign genes can be injected into the tumor, where they will merge with the fatty cell membranes of the cancer cells and dispense their alien DNA.

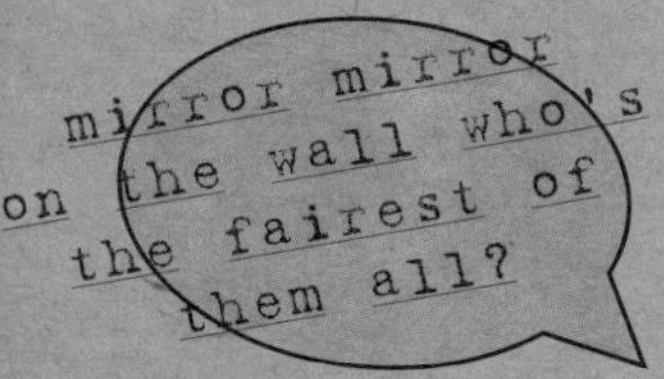

aging is a fact of life

GROWING OLD

* There aren't too many certainties in life, but aging is definitely one of them. Many of us hate the idea of growing old and are prepared to spend our hard-earned cash on "miracle" creams and potions in an effort to delay the inevitable. But if you're looking for someone or something to blame for this cruel twist of fate, then blame it on your genes.

WEAR AND TEAR

* Nobody is really sure why we age, but there are plenty of theories—and genes, almost inevitably, feature prominently in all of them. The most obvious explanation is that ***aging is simply a consequence of the accumulation over the years of mutational damage in our DNA***.

* Mitochondria are the cell's powerhouses, and damage to their DNA is the equivalent of a factory shut-down. If you compare the cells of old people with those from younger people, the older cells have many more inactive mitochondria. A gradual shut-down in energy production, caused by years of accumulated mutations, is bad news for the cell—and even worse for its owner!

Live Longer, But...

Any males out there ever considered castration? It might not improve your sex life, but it could keep you on the planet for a few extra years. There is some evidence to show that eunuchs live, on average, 15 years longer than noncastrated males. This suggests that testosterone, the male sex hormone produced in the testes, may perhaps play a part in the aging process.

* Mutational damage accumulates most rapidly in the mitochondria because they lack the enzymes that help patch up and repair damaged DNA. But it also occurs in the chromosomal DNA. Nuclear DNA does produce repair enzymes, but the genes that code for these enzymes are themselves vulnerable to mutations. Nothing is perfect in the Garden of Eden!

TAKE A TIP

* ***An alternative aging theory offered by some scientists highlights the importance of telomeres***. A telomere is just a fancy name for the tip of a chromosome. Telomeres consist of hundreds of repeats of the sequence TTAGGG. Every time a cell divides and the chromosomes are replicated, the telomere of each chromosome is shortened by a few of these repeat units. The repeat units serve as anchor points for the DNA polymerase to bind onto when it begins replicating the chromosome. The polymerase can replicate the entire chromosome except the bit that it's first bound to—so each time the cell divides, the chromosome is shortened.

* Once the telomere is shortened beyond a critical length, the cell is unable to divide any more and dies. However, the cell can be rejuvenated by adding an enzyme called TELOMERASE.

Sun-dried

If you spend too much time in the sun, you could find that your skin shows signs of premature aging—otherwise known as wrinkles and blemishes. The reason? The DNA in your skin cells does not take kindly to unremitting UV rays, and the mutational damage piles up.

sun ages the skin prematurely

Eternal youth

Telomerase is usually only found in the germ-line cells, where it acts to ensure that the next generation will inherit full-length telomeres. Some scientists have suggested (unwisely!) that telomerase could hold the secret of eternal youth.

no one knows where the rogue proteins are hiding

MAD COWS AND ENGLISHMEN

★ Until a few years ago, the scientific consensus was that all infectious diseases were caused by bacteria or viruses that carried their own genetic material. But times have changed, and infectious agents are now turning up that have neither DNA nor RNA. These are the prion proteins, rogue versions of the body's ordinary proteins that can wreak havoc on the nervous system.

GRIM NUMBERS

Nobody knows how many cows are, or have been, infected with BSE. In the last few years, over a million cows have been slaughtered in Britain alone. But some scientists estimate that as many as 50,000 people could have eaten infected meat before the disease was first diagnosed in 1986.

GOOD OR BAD

★ We all have a protein in our bodies called PrP, which attaches to the outside of nerve cells, particularly those in the brain. *Although its function is unknown, in its normal form PrP is completely harmless. But the weird thing about PrP is that it can flip from the normal "good" form into a "bad" form*. A change of shape is the only discernible difference between the two forms, but the effect of this change on the body can be catastrophic. The structural alteration to the protein messes up the nerve cells, leading to progressive neural degeneration and death.

PRION DISEASE

★ Current research suggests that PRION DISEASES can occur in at least two ways.

the meat scare still rages

The first, and more orthodox, route is through mutation. In humans, for example, CREUTZFELDT-JACOB DISEASE (CJD) is thought to be caused by a mutation in the PrP gene. The mutation increases the chances that the "good" form of the protein will flip into the "bad" form. Sheep have a similar inherited form of prion disease, called scrapie.

* The alternative, infectious, route of the disease is the one that has been making news headlines. BSE (BOVINE SPONGIFORM ENCEPHALOPATHY), or mad-cow disease, is not a "natural" inherited form of prion disease. Ground-up sheep brains were a standard constituent of animal feeds for many years; when BSE was first diagnosed in Britain in 1986, the finger of suspicion pointed toward animal feed

BSE may have been passed to humans through the consumption of infected meat

containing scrapie-infected sheep brains. Once ingested, the sheep prions were able to bind to the "good" prions in the cows' brains and twist them into the "bad" shape.

MORE BAD NEWS?

There now appears to be a new form of Creutzfeldt-Jacob disease in humans, caused by eating BSE-infected meat. The "bad" cow prions seem to be doing in humans what the "bad" sheep prions did in cows. Nobody knows how much infected meat has been consumed, or whether the U.K. is on the brink of a new epidemic. It's still thought possible that most people will be resistant to the protein-flipping antics of the mad-cow prions.

ONGOING RESEARCH

In 1997, **Stanley Prusiner**, professor of biochemistry at the University of California in San Francisco, was awarded a Nobel prize for his pioneering work on prion diseases. Because of the BSE and CJD scares, prion diseases have now become the subject of intensive research.

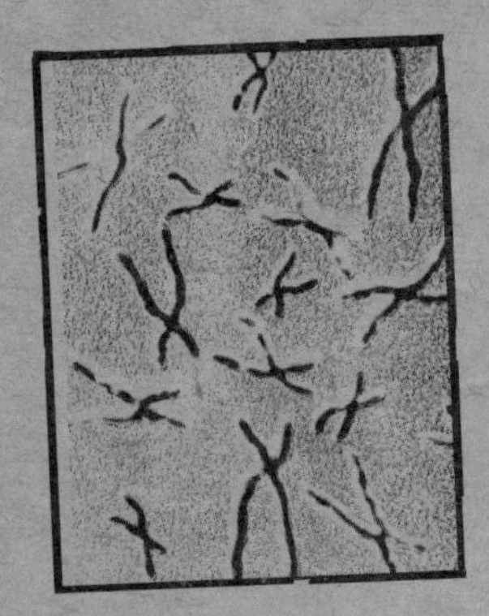

soon the genetic recipe will be known

Sequencing the genome

In part, the race to sequence the human genome has as much to do with financial gain as scientific understanding. There is a desire among some scientists and many biotechnology companies to patent human genes, particularly those responsible for disease. Disease-causing genes have great medical importance, and so could be very lucrative for drug companies and scientists alike. But the idea of owning the rights to a stretch of DNA sequence is a bizarre concept. After all, scientists don't invent genes, they merely identify them.

THE FUTURE

★ The Human Genome Project is one of the most ambitious scientific projects since the space race. Its ultimate objective is to map the positions of all the 100,000 or so human genes on the 23 chromosome pairs. This will involve working out the entire sequence of DNA letters in the human genetic recipe—all 3 billion of them.

soon we will all be mapped out

MAN MACHINE

★ Once it became possible to sequence the entire human genome from beginning to end, it was inevitable that this ambitious goal would be attempted. The task is an onerous one, but over recent years it has been made considerably easier due to significant advances in sequencing technology. Nowadays, DNA sequencing is almost fully automated. A machine, rather than a human being, does most of the work. You can stick a tube containing a bit of DNA into one end of a machine and

the database will contain all human DNA sequences

within a few hours get the sequence of DNA letters out of the other end.

* Although this system works well, it is still only possible to read about 400 letters at a time. Working every day of the year, it would take one person about 10,000 years to sequence the entire human genome. It is not surprising, then, that the Human Genome Project has been divided up between many scientists in several different countries, including Britain and the U.171S. Their aim is to finish the sequencing by the year 2005.

LONG-WINDED

* What will the complete DNA sequence tell us about ourselves? Well, initially not very much. On its own, the DNA sequence will be a pretty boring read. Containing a thousand times as many letters as Tolstoy's *War and Peace*, it will say little about where genes are or what they do. Nevertheless, the organizers of the project hope that the complete DNA sequence of the human genome will act as an invaluable reference book for the study of human genetics.

Money well spent?

Because most of the genome is made up of genetic garbage, critics of the Human Genome Project argue that most of the money will go on sequencing vast stretches of DNA with little scientific interest. But even junk DNA may turn out to have some important but as yet unknown function.

scientists are working to finish by 2005

SPEEDING UP THE SEARCH

* The search for the cystic fibrosis gene provides a good example of the way geneticists find genes from scratch. Although it was a very worthwhile scientific discovery, it was a highly time-consuming exercise, involving years of hard labor.

it all started with a pea

FUTURE DEVELOPMENTS

* From everybody's point of view, it would be useful to find some way of speeding up the search for those elusive genes—and the human-genome database may one day make that possible.

the search for genes is very slow

* ***In the future, it may be much easier and quicker to locate genes by using their protein products as the primary clues in the search.*** Take cystic fibrosis again as an example. The major symptom of the disease—an excessive build-up of thick mucus in the lungs—was traced to a single defective protein in the membrane of cells. In normal individuals this protein acts as a gateway for

the passage of chloride ions across the membrane; but in cystic fibrosis sufferers, the protein gates don't function properly. Insufficient chloride ions are able to get out of the cells, and this results in an accompanying lack of water in the mucus.

scientists are gradually unlocking the secrets of our genes

* Once the protein responsible for the disease has been identified, then a simple chemical analysis will reveal its amino-acid sequence. Once you know the amino-acid sequence, you can work backward to determine the DNA code. In practice, the existence of introns (the non-coding sequences of a gene) means that this step can only be an approximation. But the amino-acid sequence should at least enable geneticists to work out a significant proportion of a gene's DNA sequence. Geneticists will then be able to search the human-genome database for a sequence that matches their own. A sequence match will automatically reveal the exact whereabouts of the gene.

* ***This method of hunting down genes is, of course, applicable not only to the ones that cause disease but to any characteristic whose variability can be traced to differences in the structure of a coded protein.***

MISSING LINKS

As each new bit of human DNA is sequenced, it is entered into an ever-growing computer database of human DNA sequences. Eventually, the genome database will contain the complete ordered sequence of every human chromosome.

sheep first,
humans next?

HELLO DOLLY!

* Sheep don't often make the headlines. But in early 1997, one sheep in particular was front-page news. Unless you've just emerged from an extended hibernation, you can't have avoided hearing about the story of Dolly the sheep. So what was all the fuss about?

Dolly was the first mammal clone from an adult cell

What is a Clone?

*A **clone** is an identical copy. In molecular biology a clone normally refers to a copy of a gene, but it can be equally applied to an entire genetic recipe, or an entire organism*

Why a Sheep?

The cloning of Dolly was carried out by a team of scientists from the Roslin Institute in Edinburgh. A sheep was chosen for cloning because sheep, like humans, are mammals—and they are also cheap and easy to keep!

CLONING

* Dolly did not start life in the way that most animals do, through the fusion of a sperm and an egg. Dolly's entire complement of chromosomes came from the cell of a six-year-old adult sheep. In other words, her mother was also her identical sister. Dolly was the first ever clone of a mammal from an adult cell. Embryos had been cloned before, but there is a big difference between cloning from an adult cell and cloning from a cell taken from an

embryo. An early animal embryo is made up of a ball of undifferentiated cells. If you remove and isolate one of those cells, the cell has the potential to develop into an embryo on its own. It will grow and divide to form an identical clone of the embryo from which it was derived. This is why cloning animals from embryo cells is relatively straightforward. Each cell is undifferentiated and has the ability to grow into an organism in its own right.

* But as the embryo continues to grow, at a certain point the cells lose their autonomy and independence. As specific combinations of genetic switches are turned on and off, the cells become committed to following a specific path of development—to become muscle cells, skin cells, and so on.

adult cells are fully developed

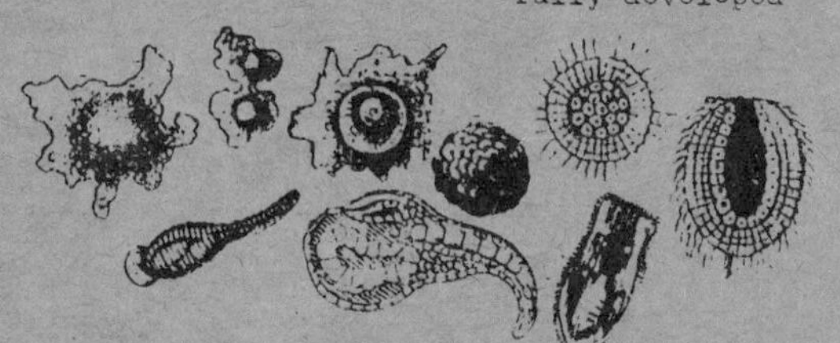

FATE

* The amazing thing about Dolly is that her genetic recipe came from the cell of a six-year-old sheep, a cell whose fate had long been decided. So the scientists who created Dolly worked out a way of overriding the rules of nature and persuading the cell's genetic recipe to regain its independence.

CLONING CLAIMS

In 1981, two scientists in Switzerland claimed that they had cloned mice from unspecialized embryo cells—the first mammals to be cloned. Later experiments raised doubts about these claims. In 1986 sheep and cows were cloned from embryo cells. This time there was no dispute.

Frog failures

In the 1960s, failed attempts to clone frogs convinced scientists that cloning from adult cells was an impossible dream. Dolly proved otherwise.

CLONING COOKERY

* After all the hype, it may come as something of a surprise to learn that Dolly is not actually a perfect clone. A perfect clone has identical nuclear DNA and mitochondrial DNA. Dolly only received her nuclear DNA from the six-year-old sheep; her mitochondrial DNA came from a different sheep altogether. Here's how it was done.

Sheepish Clue

Have you ever wondered how Dolly got her name? Here's a clue: it has something to do with udders and a certain famous namesake. Scientists like to have their little jokes.

STEP-BY-STEP CLONING

* First, a cell was taken from the udder of an adult sheep. Unlike Dolly, this six-year-old sheep doesn't seem to have been given a name, so will henceforth be referred to as Sheep No. 1. After it had been removed, the udder cell was "starved"—which somehow, in ways that are still not fully understood, made it "forget" it had ever been an udder cell and regain its youthful independence.

* Enter Sheep No. 2. An egg was taken from Sheep No. 2, and its nucleus was removed. Its mitochondrial DNA, however, was left untouched. The chromosomes were then extracted from the udder cell of Sheep No. 1 and inserted into the "empty" egg cell nucleus of Sheep No. 2.

NEW LIFE

★ The egg was then given an electric shock (the geneticist's equivalent of using a spark plug on a car) to jump-start its development, before being inserted into the womb of yet another sheep, where the embryo continued its development until birth.

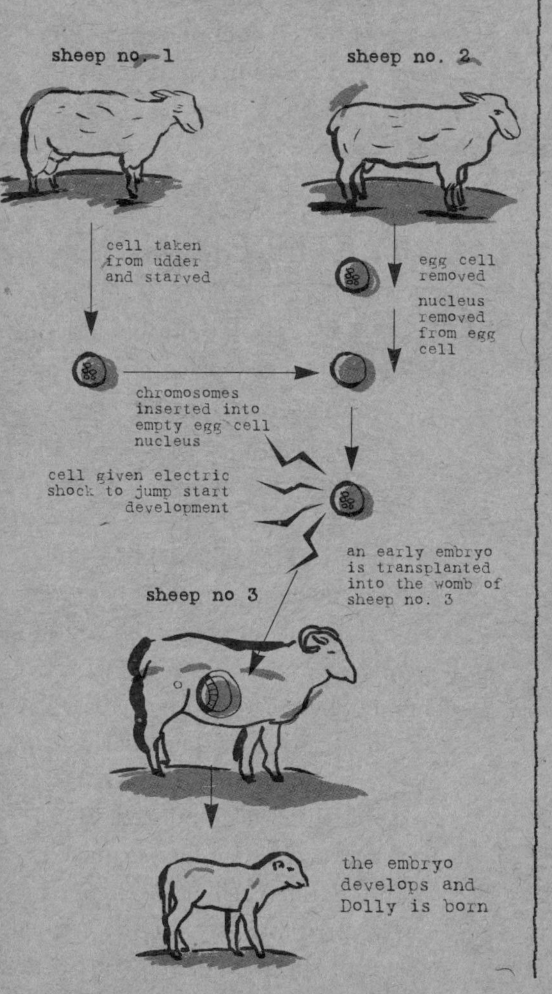

BONNIE AND DOLLY

At 4 a.m. on April 13, 1998, Dolly gave birth to Bonnie. Bonnie was conceived by conventional means, and her birth seems to provide reassurance that cloned animals, like Dolly, can develop into healthy animals capable of reproducing.

Not so easy

Although the whole procedure may sound relatively straightforward, there were some 277 failed attempts before Dolly emerged into the world. The scientists found that cloned embryos grew much faster than normal ones, and many were still-born or had to be removed by cesarean section.

sheep cloning has already produced benefits

AN OLD SHEEP IN YOUNG SHEEP'S CLOTHING

Dolly could help us to understand how aging works. Although she's only two years old, her DNA is eight years old, and researchers are watching closely for premature signs of aging.

The A to Z of Cloning

First we had Dolly, then we had Polly. What can we expect next from the world of sheep cloning? Folly, perhaps?

THE BENEFITS OF CLONING

* Researchers are not cloning sheep just for the fun of it. There are long-term goals which could one day bring real medical benefits to humans. In fact, a new sheep clone is already helping in the treatment of hemophiliacs.

POLLY POTENTIAL

* Not long after Dolly was born, a sheep named **Polly** made her debut. Unlike Dolly, ***Polly was a clone of an embryo cell, rather than an adult cell***. But the embryo was no ordinary embryo. A working copy of the human blood-clotting gene that is defective in hemophiliacs had been inserted into its genetic recipe. Polly has grown up producing human blood-clotting protein in her milk, which can be purified and then injected into hemophiliacs. And, of course, once one hemophiliac-friendly sheep has been genetically engineered, cloning makes it possible to produce many more such sheep rapidly and efficiently.

if one, why not several?

ETHICS ASIDE

* It may be only one small scientific step from cloning a sheep to cloning a human, but it's a gigantic ethical leap for mankind. Nevertheless, leaving aside the ethical considerations for a moment, there is a strong scientific and medical case to be made for developing human-cloning technology.

the ethical problems of cloning are considerable

BENEFITS

* One possible application is in the treatment of a rare disease caused by a defective mitochondrial gene, which can result in blindness and epilepsy. After the creation of an embryo by *in vitro* fertilization, the nucleus would be removed, leaving the defective mitochondria behind. This would then be inserted into a donor egg, from which the nucleus has been removed, but not the healthy mitochondria.

Cloning and the Law

Human cloning has been banned in Britain. However, there are signs that the ban may be lifted if a strong enough case is made for the medical benefits cloning could provide.

NEW POSSIBILITIES

Cloning could also be used to provide more human embryos for medical research. In Britain, embryos can be used for research purposes for the first 14 days after conception. Cloning would allow many genetically identical copies to be made, which could create new possibilities for the study of human development and disease.

would we really want a clone of Joseph Stalin?

TO CLONE OR NOT TO CLONE

* For every claim of the positive medical benefits of cloning, there is an apocalyptic counter-claim. Scientists are accused of playing God, of interfering with nature. And what if the technology fell into the wrong hands? The media is full of sensationalist scenarios of human clones being used as factories for the growth and sale of human organs on the black market, or of ego-crazed dictators cloning themselves to prolong their regime.

THE NOVELTY FACTOR

* New technologies are often distrusted, particularly biological ones. When *in vitro* fertilization was first introduced in the 1970s, many people dismissed it as "unnatural." Yet since then, IVF has enabled thousands of infertile couples to have a child; today nobody thinks twice about it. It has become just another medical treatment. Cloning could also provide medical benefits. ***Sure, it's unnatural. But then so is most modern medicine.***

organ cloning could become big business

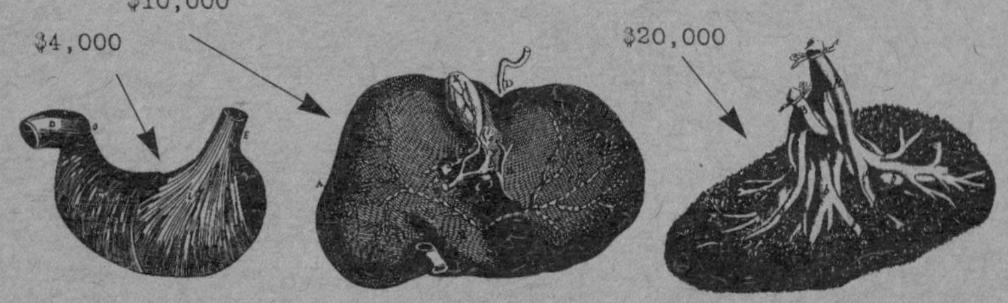

MORALS AND MATERIALISTS

★ ***But what of the accusation that cloning could be misused, that material considerations could displace the rights of the individual? This is a serious moral concern***. Most sane individuals would feel uneasy with the notion of humans as growth factories, as the organic equivalent of a back-up computer disk. Of course, if there is money to be made, then undoubtedly someone will try it. But is the risk of the technology being misused an adequate argument for never using it?

are scientists playing God?

HUMAN DOLLY

★ Most fears about cloning seem to be concerned with the specter of a human Dolly, of taking a cell from someone and using it to make an identical genetic copy of that person. Early in 1998, ***Dr. Richard Seed***, an American scientist based in Chicago, made an announcement stating that he hoped to set up a clinic for just such a purpose. In the future, he wants to offer a cloning service for anyone who feels they need a copy of themselves. One thing is certain—he would have no shortage of weirdos wanting to make use of his service!

SHADES OF FRANKENSTEIN?

Even if someone does develop the technology to do it, cloning an adult from one of their cells could have unpredictable consequences. A cell taken from an adult may have accumulated all sorts of DNA damage. Making the DNA of that cell the genetic recipe of a new life could be the genetic equivalent of Russian roulette.

GENETIC MYTHS

★ Cloning has certainly caught the public's imagination. But perhaps the fear and fascination of cloning reflects the fact that society is thinking too much about genes, and not enough about the other things that make us what we are. The idea of having an identical "you" may be seductive or appalling. Either way, it defies reality. Having an identical genetic recipe doesn't make two people exactly the same. After all, human clones already exist. They're called identical twins.

how would someone with Hitler's genetic recipe turn out today?

GENE MACHINES?

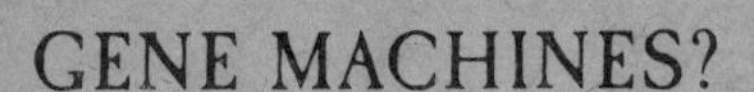

★ Identical twins have similar features, and they may even have similar likes and dislikes. But usually, in addition to sharing genes, they share the same home

identical twins share the same genes and features

TWINS

Identical twins arise when the cells of a two-cell embryo separate from one another and go on to develop independently as two embryos. Fraternal twins are simply the result of two different eggs being fertilized simultaneously by two different sperm. Genetically speaking, fraternal twins are no more similar than brothers and sisters born separately.

environment while they are growing up. Although genes undoubtedly have an important influence on physical development and appearance, and may even have a broad influence on personality, they do not work in isolation. Human beings are not mere gene machines. Our environment—the circumstances in which we grow up, our diet, lifestyle, and unique set of social and family experiences—also has an important part to play in how we are as individuals. * If some mad scientist found a perfectly preserved copy of Hitler's genetic recipe and made a cloned individual from it, you wouldn't end up with another Hitler. ***This is because memory and experience are not preserved in DNA***. Hitler was a product of his times and of his own unique set of social experiences, not just a product of his genes.

PIPE DREAM

* Thankfully, the technology required to re-create a long-dead person is a pure science fiction pipe dream—for the time being, at least. If Hitler's genetic recipe was ever brought back to life, then, certainly, the person created might look a bit like Hitler—although in the 21st century he would probably ditch that silly moustache. But he'd be just as likely to end up as a social worker, a PE teacher, or a chef, as a Fascist dictator. Genes aren't everything.

Mother Teresa

Missed Opportunity

Dr. Richard Seed, Chicago's self-professed "Dr. Clone," once remarked that he wished he had taken a blood sample from Mother Theresa before her death, so that he could have created a cloned replica. His comment revealed a fundamental misunderstanding of genetics.

fraternal twins are not genetically identical

GENES AND CAKES

★ The relationship between you and your genes is a bit like that between a cake and the recipe required to bake it. The cake recipe does not make the cake on its own. It is simply a set of instructions or guidelines which, if followed correctly, should produce something resembling a cake. As anyone who has dabbled in the kitchen knows only too well, changing the ingredients, or the temperature of the oven, can have unforeseen consequences.

LIFESTYLE

★ *Likewise, our genes are simply a recipe for our bodies. They do not work independently of the conditions in which we develop*. Take human weight, for example. There is no doubt that genes have some influence over our weight. Even if everyone always ate the same amount of food, our weights would not all be the same. Because of their genes, some people burn off fat more quickly than others. You could say that some people were genetically predisposed to be fat or thin—but that doesn't mean your waistline is determined exclusively by your genes.

eating too much cake is more likely to make you fat than your genetic disposition alone

* People who gorge themselves on burgers and soda will become fat, no matter what their genetic recipes say, since behavior and environment can override the influence of genes. In many cases, genes simply mark out the boundaries within which development can take place—and sometimes those boundaries can be very broad indeed.

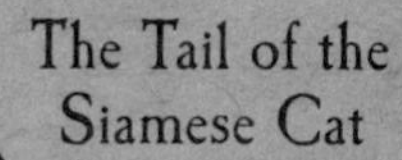

lifestyle can influence genes

CHEATING GENES

* Even disease-causing genes are not always as deterministic as they may first seem. People who have the genetic "disease" phenylketonuria, for example, are unable to metabolize the amino acid phenylalanine. But if they avoid phenylalanine in their diet, then effectively the disease does not exist.

* Conversely, if everyone smoked, then lung cancer would be regarded as a genetic disease, because people vary in their genetic susceptibility to cancer. ***Often, genes and their effects can be seen only in the context of their environment.***

The Tail of the Siamese Cat

Siamese cats provide an illuminating example of the way in which gene products can interact with the environment. An enzyme that controls the synthesis of the black pigment in the cat's fur is sensitive to heat, and is only active at cool temperatures. So black fur tends to be restricted to the cooler parts of the body: the ears, tail, nose, and feet. Sticking a young kitten in the fridge would be ethically unacceptable—but it would result in a completely black cat!

GENES, ENVIRONMENT, AND TWINS

★ Studying twins provides a potentially informative way for biologists to assess the relative influence of genes and environment on human characteristics. Yet such studies have often been shrouded in controversy and scandal. Nowhere is this more true than in studies of the genetic basis of IQ.

studies into twins have been inconclusive

BLACK AND WHITE?

In the U.S., tests have suggested that the average IQ of black people is lower than that of whites. In 1994, Richard Herrnstein and Charles Murray's controversial book *The Bell Curve* asserted that this difference was due primarily to genetics rather than environment. Their ideas, echoing those of Francis Galton 100 years ago, have been condemned and severely criticized by many scientists.

SHARED LIVES

★ Identical twins share the same genes. So if a particular characteristic is influenced by genes alone, then both twins should have it. The trouble is that because twins normally share the same family environment, similarities between them could be caused either by similar genes or by their shared environment.

SEPARATED AT BIRTH

★ To distinguish between these two possibilities, geneticists look at twins that have been separated at birth and have grown up with entirely different families, usually in different locations. If they are still similar, it suggests that the similarity is caused by genes. Although it sounds fairly easy, carrying out this sort of study is fraught with experimental difficulties, not the least of which is finding enough pairs of twins of the same age to compare.

* Several twin studies have looked at the influence of genes on IQ. One study came up with the result that 80% of the variation in human IQ scores was genetically based. In other words, a person's IQ was influenced primarily by the genes they inherited from their parents—and environment and education exerted only a minor influence.

* It was later revealed that the entire published study had been fabricated, including the names of its authors. But if it had been genuine, one can scarcely imagine what the social and political ramifications of such a result would have been.

how are differences between twins to be explained?

can intelligence be acquired?

Testing Intelligence

If IQ does have a large genetic component, and many scientists believe that it does, there is still the contentious question of whether IQ is an accurate measure of intelligence, or simply a reflection of a person's ability to perform a specific and limited set of cognitive tasks.

is IQ just a measure of ability to perform a limited set of tasks?

is crime simply a genetic defect?

GENES AND BEHAVIOR

* The idea that genes could influence behavior and personality stirs up strong emotions. Sexuality, aggression, anxiety, criminality, and alcoholism, to name but a few, have all been linked to genes. But this doesn't mean that we are helpless prisoners of our genes—although society often seems to take a different view.

X MARKS THE SPOT

* In 1993, **Dr. Dean Hamer** of the U.S. National Cancer Institute published a study which showed that homosexuality in men tends to run in families. Hamer found that the incidence of homosexuality among the brothers of men known to be gay was much higher than you would expect by chance alone. This suggested an inherited component to sexual orientation.

* What's more, the maternal uncles and maternal male cousins of gay men also showed a higher incidence of homosexuality than expected, indicating that the genetic influence was associated with the X chromosome.

* Always on the lookout for a good headline, the world's press responded to Hamer's study by announcing the discovery of the "gay gene." Hamer himself

"We have not found the gene, which we don't think exists, for sexual orientation."

DEAN HAMER, 1993

EROGENOUS ZONE

By 1995, Hamer had identified a region of the X chromosome that seemed to be associated with sexual orientation. The region was sufficiently large to include tens or perhaps hundreds of different genes.

was quick to dismiss the notion, stressing that all he had uncovered was ***evidence of a genetic component to sexual orientation in some men.***

BEHAVIOR

★ Human behaviors are complex things and are likely to be affected by a multitude of factors. It seems entirely plausible that genes could be one of the factors that affect our sexuality, as well as many other human behaviors. But when it comes to reports of scientific research, there is always a danger that society will adopt an oversimplistic view, and that some unscrupulous people will misrepresent that research for their own political ends.

Discriminatory tests

After the "gay gene" furor, there was considerable concern that the military and insurance companies might develop a diagnostic test to discriminate against homosexuals. As yet, these fears have not been realized.

GENETIC DISCRIMINATION

★ As the media response to Hamer's research shows, new discoveries regarding genes and behavior tend to be taken as evidence that behavior is determined by genes. The fear is that genetic studies of behavior could pave the way for genetic testing and genetic discrimination. In Britain, for example, the *Daily Mail* summed up Hamer's discovery with the headline, "Abortion hope after 'gay genes' finding."

GAMBLING WITH GENES

there's an element of chance with many genetic diseases

* In the future, genetic knowledge is going to present us with some difficult choices. There are already signs that people are making use of information about their own genetic recipes to plan ahead. And even if you'd rather remain oblivious to what your genes might have in store for you, you can be certain that there are plenty of people out there who are dying to know.

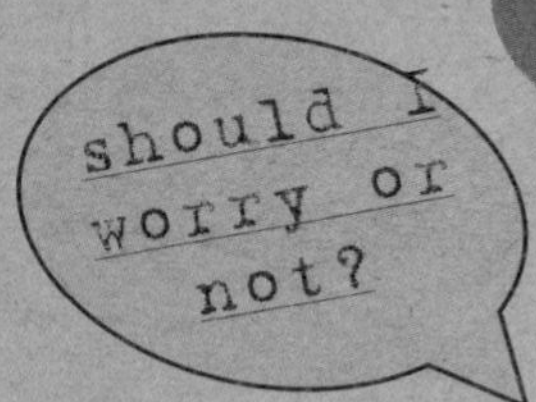

some people don't want to know their future

RECIPES AND REVELATIONS

* In the case of many genetic diseases, genes carry a clear message. If you have the gene, you will get the disease. But scientists are beginning to discover other genes that carry a vaguer message. These genes merely suggest the future, rather than predicting it with absolute certainty. They increase the odds that a person will develop cancer, have a heart attack, or develop any number of other diseases. Environmental effects also come into play—such as how much you smoke, what you eat, and how much you drink. But for most diseases, genetics ensures that we don't all start out with a level playing field.

* In 1996, there was an interesting case of a woman in Britain who decided to have

her breasts surgically removed. She had no symptoms of any illness. Her action was prompted by a genetic test that revealed she had a very high risk of developing breast cancer. There was a history of breast cancer in her family, and her own mother had died of it in middle age. Of course, there was no absolute certainty that the woman would have developed breast cancer. But she had used the information in her own genetic recipe to give herself full protection for the future.

other people want to know what is in store for them

* It's understandable that many people would rather not delve too deeply into their genetic recipes. But insurance companies certainly would. Insurance companies are always looking for ways of identifying high risk groups. If your genes, through no fault of your own, make you a high risk, then you can be sure that those oh-so-friendly insurance folk will want to know about it. ***Privacy of genetic information is currently one of the most hotly debated and contentious issues around.***

RISK ASSESSMENT

Insurance companies may not have it all their own way, even if they do gain access to the results of genetic tests. Though tests may identify individuals as belonging to high-risk groups, they may also identify a person as extremely low risk. In these latter cases, genetic testing could potentially reduce the need for any health insurance at all.

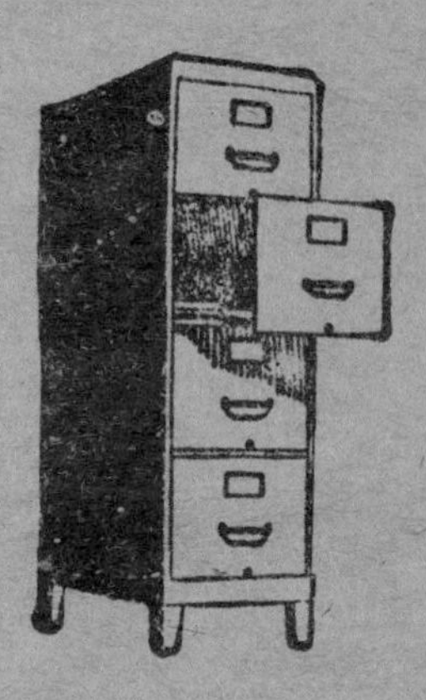

insurance companies would like to file away our genetic details

FREEDOM OF CHOICE

* The idea of choosing a child based on its genetic characteristics isn't a new one. In parts of India and China, for example, there is a long tradition of preferentially aborting female fetuses. But as new genetic tests become available, the range of choices open to parents may go way beyond whether you want a boy or a girl.

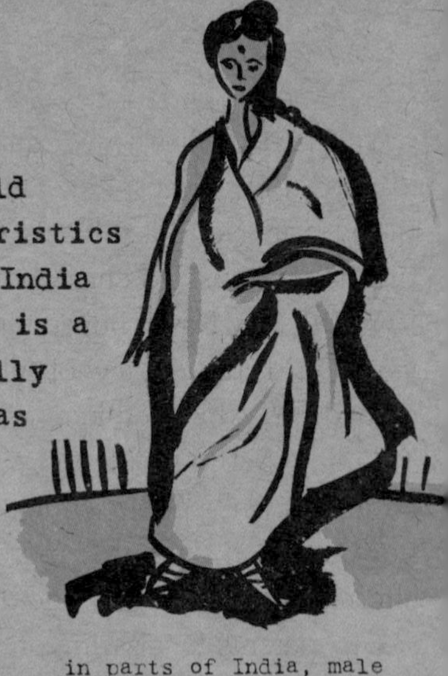

in parts of India, male children are preferred

CHOOSING CHILDREN

* In the future, genetics will continue to plow its way through an ethical minefield, constantly confronting us with new moral choices and difficult decisions. As more and more genes are uncovered, genetic testing is likely to extend beyond the diagnosis of the most serious genetic diseases.

* It's even possible that parents of the future will be able to choose the characteristics of their children in the same way they'd choose a car, making up their

will we be able to choose our child's characteristics?

own genetic recipe from a million different options. Brown eyes? No. Blue eyes? Yes. Blond hair? No. Black hair? That sounds nice! Some chiseled cheek bones, straight from the new David Bowie range of genetic variation? Perhaps a few of Albert Einstein's cloned genes inserted here and there? That'll be $20,000, please!

★ If that isn't spooky enough for you, how about an altogether more surreal alternative? Will a child one day be able to sue its parents for passing on a defective gene—a case of genetic negligence—if the technology was available to detect and correct the defect before birth? Might prospective parents be legally required to pass a genetic test before having children?

MORAL MINEFIELD AHEAD

★ All this wild speculation is an indirect way of saying that ***in the future genetics is going to present society and individuals with choices and decisions that we have never had to make before***. Should we let the results of genetic tests decide the fate of an embryo? If so, what should we test for? Just the most serious genetic diseases, or any of the thousands of other human characteristics? Who should decide whether tests should be done? The individual, a doctor, the state? Genetic engineering and genetic testing seem to offer many new and exciting medical benefits, but ***without careful legislation we could end up with moral mayhem***.

in parts of India, girls aren't considered to be economically viable

Tilting the Balance

In certain parts of India, parents prefer sons to daughters because daughters are much more expensive. When they grow up and marry, the family is expected to provide a dowry. As a result, in some states there are now 20% more men than women.

FREEMARKET FATE

★ Genetics has come an awfully long way since its humble beginnings amid the pea plants of an Austrian monk. It is truly amazing what has been achieved in a little over a hundred years. Its short history has been a mixed one, sometimes thrilling, sometimes cruel, and always with the whiff of controversy.

what can we expect in the future?

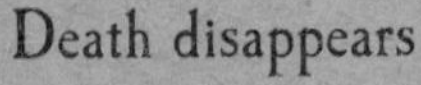

Death disappears

If science ever comes to grips with how we age, then perhaps one day growing old and even death itself may become a thing of the past.

what more is there to discover about our genetic history

THE SHOCK OF THE NEW

★ What can we expect from genetics in the future? In a few years the Human Genome Project will have recorded the entire DNA sequence of a human being. Will it tell us what it is that makes us human, what it is that sets us apart from our evolutionary relatives?

THE FINAL FRONTIER

★ ***The Human Genome Project will unmask the genes for the thousands of characteristics that help to create the huge diversity of human life.*** It could help us understand one of life's great mysteries—how a fertilized egg develops into something as mind-bogglingly complex as a human being—and it will present new opportunities for the treatment and cure of all genetic diseases.

EUGENICS REVISITED?

★ But the era of freemarket fate may soon be upon us. As new genetic tests become available, the specter of eugenics will once again rear its head. Should we allow the results of genetic tests to decide the fate of our offspring? Prenatal testing for genetic diseases already allows us to exercise some level of choice. Will this be extended to other genetic characteristics? Who will benefit most from the new technologies, and will they actually improve the quality of our lives?

Ironic peas

In some ways, it is ironic that the man who started this scientific and conceptual revolution was himself deeply religious. But when Gregor Mendel began his work on peas, he sowed the seeds of a science that shows no signs of slowing down.

JUST THE START

★ We are fascinated by the idea that our fates can be determined by our genes. For thousands of years, people lived their lives in the belief that their future was uncertain and unknowable. You might live to be a hundred, or you might die young of some horrible disease. You put your trust in luck, or perhaps in God. But knowledge about our genes is beginning to change all that.

★ Peering into our genetic recipes is like peering into a crystal ball. The difference is that ***modern genetics not only offers a glimpse of the future, it provides a way of changing it***. So much for the last hundred years. Roll on the next hundred!

Mendel probably didn't realize where his investigations would lead